Build Stunning Real-time VFX with Unreal Engine 5

Start your journey into Unreal particle systems to create realistic visual effects using Niagara

Hrishikesh Andurlekar

BIRMINGHAM—MUMBAI

Build Stunning Real-time VFX with Unreal Engine 5

Group Product Manager: Rohit Rajkumar

Publishing Product Manager: Vaideeshwari Muralikrishnan

Senior Content Development Editor: Debolina Acharyya

Technical Editor: Saurabh Kadave

Copy Editor: Safis Editing

Project Coordinator: Manthan Patel

Proofreader: Safis Editing

Indexer: Hemangini Bari

Production Designer: Nilesh Mohite

Marketing Coordinators: Nivedita Pandey, Namita Velgekar, and Anamika Singh

First published: April 2023

Production reference: 1270423

Published by Packt Publishing Ltd.

Livery Place

35 Livery Street

Birmingham

B3 2PB, UK.

ISBN 978-1-80107-241-0

www.packtpub.com

To my mentor, Shri Sharad Londhe, for all the different perspectives that helped open my mind; my parents, for their support; my wife, Dr. Nilangi Andurlekar, for tolerating me as I wrote this book; and my friends and all the students who unwittingly were the testers for the content of this book.

– Hrishikesh Andurlekar

Foreword

It is my pleasure to write a foreword for this book by Hrishikesh Andurlekar. I have known him for more than 18 years in different roles, from a CG supervisor to a real-time technology expert. Real-time technology has matured over the years and has moved from games to VFX. I have had the good fortune of working with computer graphics in movies as well as interactive graphics for games and virtual reality, and appreciate the overlap of the skillsets in the current games and VFX industry.

After enabling the creation of stunning effects in games, real-time game engines are revolutionizing the way visual effects are created for movies. With the power and speed of modern game engines such as Unreal, VFX artists can now create stunning visual effects in real time, eliminating the need for lengthy render times and costly post-production processes.

This technology has been used to create everything from realistic virtual sets to intricate particle effects, allowing filmmakers to create immersive and realistic worlds on a scale that was previously impossible. By using real-time game engines, VFX artists can see the results of their work immediately, making it easier to iterate and refine their designs.

At Annapurna Studios, we have leveraged real-time technology for movie production, having set up the largest LED volume in India. As our teams started adopting the technology, we felt the need for good training content that explains the basics of the technology, which would lay a solid foundation for our artists.

This book does a fantastic job of helping anybody new to real-time VFX to get started without being overwhelmed. It helps you work your way up and get comfortable with the tool to an extent that you start creating production-ready content. Niagara can be a bit overwhelming, even to artists who are familiar with Unreal, and this book holds your hand through the fundamentals. Hrishikesh not only demystifies the interface but also touches upon advanced concepts such as custom modules and event handlers.

Unreal Engine offers ways to enable directors to experiment on their sets in real time, and this aspect of the workflow is often overlooked by newer entrants in the virtual production field. This is possible through blueprints, and this book also introduces you to workflows where you can tweak particle systems at runtime for games or virtual production studios.

I appreciate the effort Hrishikesh has taken to encapsulate what is a pretty challenging topic into a step-by-step, easily digestible journey into learning particle effects in Niagara. I wish him all the success in the world with many more books to come.

You are in for an awesome journey into Niagara!

P N Swathi

Virtual Production Producer, AGM Annapurna Studios, Hyderabad

Contributors

About the author

Hrishikesh Andurlekar is an Unreal authorized trainer and founder of TPlusPlus Interactive, a studio delivering bespoke interactive content, including white-labeled games and simulations. TPlusPlus is also an Unreal authorized training center. Hrishikesh is a graduate of mechanical engineering from Mumbai University and has worked as a CG supervisor on movies with traditional VFX/CG pipelines with major studios before founding TPlusPlus. He has extensive experience in Unreal Engine, Unity, and Godot. He is also a speaker and a consultant on all things interactive.

About the reviewers

Benjamin Foo is a 3D generalist/Unreal artist who currently focuses on real-time rendering and virtual production. He has 12 years of experience working in broadcast branding, commercials, and film/TV VFX.

Some of the clients Benjamin has worked with include Petronas, Netflix, Astro, Samsung, and Ikea. He is also an avid digital sculptor who goes by the online handle of Jim Banne (Artstation), and he is also a real-time/Unreal Engine mentor at Mastered UK.

Benjamin currently resides in Singapore and reads way too much sci-fi in his free time.

Tuomo Taivainen is an experienced game programmer who has developed gameplay, background systems, AI, and UI in both blueprints and C++ for Unreal Engine 4 and Unreal Engine 5 on projects produced by Kerberos Productions. Tuomo is also a community member and teaching assistant for GameDev.tv.

Naser Eslami is a real-time visual effect artist. He started his career in 2012 as a visual effect artist for animation and commercial TV. In 2016, he became interested in the game industry and real-time visual effects in Unreal Engine. He developed his first game in 2013, and following that, for a few years, he used Unreal Engine for commercial TV and short animation. Today, he is a content creator for the Unreal Engine marketplace, and he loves to create stunning real-time visual effects for Unreal Engine.

Zubaida Nila is an avid academic researcher in extended reality virtual production and visual effects who has a strong background in leading and collaborating with creative industry professionals. She is the first Unreal Engine fellow based in Malaysia who has gone through intensive training with Epic Games under the women creators program. She conducts Unreal workshops at creative industry events and universities, with hundreds of participants joining her speaker sessions and having hands-on experience using Unreal Engine 5, which mostly focuses on lighting and cinematography. Besides running technical workshops, she also seizes opportunities to work on local films and commercials as a visual effects artist. Zubaida is currently working on her thesis for her master's and is on the journey to becoming an authorized Unreal trainer.

Table of Contents

Part 2: Dive Deeper into Niagara for VFX

6

Exploring Dynamic Inputs   101

7

Creating Custom Niagara Modules   115

8

Local Modules and Versioning   153

9

Events and Event Handlers   185

Preface

Unreal Engine's Niagara is a powerful visual effects system developed by Epic Games for their Unreal Engine. It allows developers and artists to create stunning and complex particle effects and simulations in real time for use as high-quality visuals in games, films, and other interactive experiences. The learning curve is, however, very steep, and it is very difficult for beginners to get started with Niagara. This book addresses this issue and gives a gentle but detailed introduction to Niagara.

Who this book is for

This book is aimed at beginners who want to learn about Unreal Engine's Niagara, a powerful visual effects tool used for creating complex particle simulations in real time. Whether you are a game developer, a visual effects artist, or a hobbyist, this book provides a comprehensive introduction to the Niagara system, its features, and its functionality. With clear explanations, practical examples, and step-by-step guidance, this book will empower you to create stunning particle simulations and visual effects in Unreal Engine.

What this book covers

Chapter 1, *Getting Started with Unreal Engine Particle System Frameworks*, gives a little history of particle systems in Unreal Engine.

Chapter 2, *Understanding Particle System Concepts*, helps you learn about foundational particle system concepts.

Chapter 3, *Exploring Niagara Concepts and Architecture*, teaches Niagara-specific concepts and provides an overview of its architecture, relevant terminology, and workflow.

Chapter 4, *Building Our First Niagara System*, gives an introduction to the UI and helps you create your first system.

Chapter 5, *Diving into Emitter-System Overrides*, takes a look at module and parameter overrides and workflow tips.

Chapter 6, *Exploring Dynamic Inputs*, helps you learn how to use dynamic inputs to extend parameter inputs and complex behaviors.

Chapter 7, *Creating Custom Niagara Modules*, explores extending the power of Niagara with custom-built modules.

Chapter 8, Local Modules and Versioning, demonstrates quick and dirty module development techniques and how to keep track of versions.

Chapter 9, Events and Event Handlers, examines the interaction between emitters using events and event handlers.

Chapter 10, Debugging Workflow in Niagara, delves into working with the Debugger panel, debug drawing, and debug console commands.

Chapter 11, Controlling Niagara Particles Using Blueprints, teaches you how to make an easy-to-use asset embedding a Niagara system in a blueprint and control the Niagara system through public variables.

To get the most out of this book

Before diving into Unreal Engine's Niagara, it's recommended that you have a basic understanding of the Unreal Engine editor. Additionally, familiarity with visual scripting languages such as Blueprint or similar programming languages will be helpful but is not required.

It's also beneficial to have a general understanding of computer graphics and the concepts behind particle systems, such as emitters, particle lifetimes, and particle attributes. Familiarity with vector math and basic physics concepts can also be useful when working with particle simulations. Some of these basics are covered in the initial chapters.

Software/hardware covered in the book	Operating system requirements
Unreal Engine 5.1 – Niagara	Windows, macOS, or Linux

The recommended OS is Windows 11.

All the exercises in the book are available in a project posted on the GitHub link given in the next section. If your Niagara systems are not working as intended, please download the project file to find examples of working Niagara systems and figure out what you may have overlooked. If you are using a more recent version of Unreal than 5.1.1, please double-check for any errors that may crop up in the project during the project upgrade.

Download the example code files

You can download an Unreal project file for this book from GitHub at https://github.com/PacktPublishing/Build-Stunning-Real-time-VFX-with-Unreal-Engine-5 . If there's an update to the project, it will be updated in the GitHub repository.

We also have other code bundles from our rich catalog of books and videos available at https://github.com/PacktPublishing/. Check them out!

Download the color images

We also provide a PDF file that has color images of the screenshots and diagrams used in this book. You can download it here: `https://packt.link/jM6sa`.

Conventions used

There are a number of text conventions used throughout this book.

Bold: Indicates a new term, an important word, or words that you see onscreen. For instance, words in menus or dialog boxes appear in **bold**. Here is an example: "These include the **Beam** type, **GPU sprites** type, **Mesh** type, and **Ribbon Data** type."

> **Tips or important notes**
> Appear like this.

Get in touch

Feedback from our readers is always welcome.

General feedback: If you have questions about any aspect of this book, email us at `customercare@packtpub.com` and mention the book title in the subject of your message.

Errata: Although we have taken every care to ensure the accuracy of our content, mistakes do happen. If you have found a mistake in this book, we would be grateful if you would report this to us. Please visit `www.packtpub.com/support/errata` and fill in the form.

Piracy: If you come across any illegal copies of our works in any form on the internet, we would be grateful if you would provide us with the location address or website name. Please contact us at `copyright@packt.com` with a link to the material.

If you are interested in becoming an author: If there is a topic that you have expertise in and you are interested in either writing or contributing to a book, please visit `authors.packtpub.com`.

Share Your Thoughts

Once you've read, we'd love to hear your thoughts! Scan the QR code below to go straight to the Amazon review page for this book and share your feedback.

https://packt.link/r/1801072418

Your review is important to us and the tech community and will help us make sure we're delivering excellent quality content.

Download a free PDF copy of this book

Thanks for purchasing this book!

Do you like to read on the go but are unable to carry your print books everywhere?

Is your eBook purchase not compatible with the device of your choice?

Don't worry, now with every Packt book you get a DRM-free PDF version of that book at no cost.

Read anywhere, any place, on any device. Search, copy, and paste code from your favorite technical books directly into your application.

The perks don't stop there, you can get exclusive access to discounts, newsletters, and great free content in your inbox daily

Follow these simple steps to get the benefits:

1. Scan the QR code or visit the link below

https://packt.link/free-ebook/9781801072410

2. Submit your proof of purchase
3. That's it! We'll send your free PDF and other benefits to your email directly

Part 1: Introduction to Niagara and Particle Systems in Unreal Engine 5

The objective of this section is to familiarize you with the basics of particle systems and introduce you to the Niagara user interface. It walks you through creating your first Niagara system and covers the system hierarchy and the workflow around it.

This section comprises the following chapters:

- *Chapter 1, Getting Started with Unreal Engine Particle System Frameworks*
- *Chapter 2, Understanding Particle System Concepts*
- *Chapter 3, Exploring Niagara Concepts and Architecture*
- *Chapter 4, Building Our First Niagara System*
- *Chapter 5, Diving into Emitter-System Overrides*

1

Getting Started with Unreal Engine Particle System Frameworks

Unreal Engine's particle system, called Niagara, is a powerful tool for creating stunning, realistic special effects in games and other interactive applications. It allows developers to create and manipulate a wide range of particle effects, such as fire, smoke, rain, snow, and more. The particle system is highly customizable, with a wide range of settings that can be adjusted to create unique effects. It is optimized for real-time performance, making it a popular choice for game developers who want to add eye-catching visual effects to their games.

Niagara replaced Cascade as Unreal Engine's particle system because it offered several significant improvements over its predecessor. Niagara was designed to be more flexible, scalable, and performance-friendly, making it a better fit for the demands of modern game development.

Some of the key features that set Niagara apart from Cascade include a more modern and user-friendly interface, improved particle simulation capabilities, and better performance and scalability. Niagara also allows developers to create and manage particle effects using either a visual interface or code, making it a versatile tool for a wide range of use cases.

In addition, Niagara was designed to be more flexible and extensible, making it easier for developers to create custom particle effects and incorporate them into their projects. This has helped make Niagara one of the most popular and widely used particle systems in the game development industry today.

We will begin our journey into Unreal particle systems with an overview of the particle system modules in Unreal Engine. There have been major changes in the particle system workflow in Unreal as we've moved from the older Cascade particle system to the Niagara particle system over the last few versions. Unreal Engine 5 continues to support the Cascade particle system, and though we do not expect to create new assets in Cascade, we will familiarize ourselves with Cascade in this chapter in case we need to support older projects made with Cascade.

We will discuss the changes Niagara brings and learn about the features expected in the future. We will also dive into the reasons behind Niagara's development and end the chapter with some interesting use cases for Niagara.

This chapter will cover the following topics:

- Particle systems in Unreal
- The Cascade particle system
- The reasons behind Niagara's development
- Use cases for Niagara

Technical requirements

For this chapter, you need to have access to a machine capable of running Unreal Engine 5. We are going to use the default assets, which should be available with your installation of Unreal Engine.

Here are the steps to install Unreal Engine using the Epic Games Launcher:

1. **Download the Epic Games Launcher**: Visit the Unreal Engine website (`https://www.unrealengine.com/en-US/download`) and click on the **Download** button. After downloading the installation file, double-click on it to install the Epic Games Launcher.

2. **Sign in or create an account**: If you already have an Unreal Engine account, sign in. If you don't, create a new account.

3. **Launch the Epic Games Launcher**: Once you've signed in, launch the Epic Games Launcher.

4. **Install Unreal Engine**: In the Epic Games Launcher, click on the **Library** tab, find **Unreal Engine** in the list, and click **Install** next to it. The installation process will begin, and you can monitor its progress in the **Downloads** tab. Make sure that you have at least 80 GB of disk space as the full Unreal Engine installation may take up to 60 GB.

5. **Launch Unreal Engine**: Once the installation is complete, you can launch Unreal Engine by clicking on the **Launch** button in the Epic Games Launcher.

And that's it! You've now successfully installed Unreal Engine using the Epic Games Launcher.

These are the recommended system configuration requirements:

- Windows 10 (64-bit, version 20H2)
- 32 GB RAM
- 256 GB SSD (OS drive)
- 2 TB SSD (data drive)
- NVIDIA GeForce RTX 2080 SUPER

AMD Ryzen 7 5800H and above (Intel Core i9, 10th generation and above)

You can find the project we worked on in this book here:

```
https://github.com/PacktPublishing/Build-Stunning-Real-time-VFX-
with-Unreal-Engine-5
```

Let us now understand how particle systems are implemented in Unreal Engine.

Particle systems in Unreal

Particles are a bunch of assets, such as images, meshes, lights, and even fully rigged characters, which are managed by a *particle system*. The particle system enables us to manage a huge number of these elements and attach logic and behavior to them.

The term *particle system* was coined in *1982* by *William T. Reeves*, a researcher at *Lucasfilm Ltd.* while working on *Star Trek II: The Wrath of Khan*. He was developing an effect for the film where a planet was terraformed. To show this terraforming, a visual effect called the *Genesis Effect* was created where a firewall ripples across a whole planet. You can watch it here: `https://www.youtube.com/watch?v=52XlyMbxxh8`.

The term *particle system* was coined for the effect shot called the *Genesis Effect*.

While each particle in a particle system is a discrete entity, it's the combined effect of all the particles in the particle system together that creates the impression of a bigger entity, such as an exploding fireball or a fireworks effect.

A game will need a particle system to show various effects such as fire, smoke, or steam. Unreal Engine previously featured a tool called **Cascade** to create particle system effects starting from the UE3/**Unreal Development Kit (UDK)** days. **Cascade** was also available in *Unreal Engine 4*. In UE 4.20, Epic introduced the new **Niagara Fx** system to replace **Cascade** as a beta version plugin, which was not enabled by default. In later versions of Unreal Engine, Niagara came enabled by default as an option along with **Cascade**. The user interface continued to give priority to **Cascade** as the primary particle effects creation tool to ease the transition to **Niagara** gradually. This changed in UE5 where the primary method of creating particles is **Niagara** and **Cascade** exists only to support legacy projects containing Cascade particle effects.

Figure 1.1: Creating a Niagara System using the right-click pop-up menu in the Content Browser

There's also a plugin called **Cascade To Niagara converter**, which can help you convert the majority of Cascade systems into Niagara Systems. It contains a **Blueprint Function Library** and some Python scripting to help with the conversion. You can enable the plugin in the **Plugins Browser** tab in Unreal 5. The **Plugins Browser** tab can be opened by clicking on **Edit > Plugins** in the menu bar.

Figure 1.2: The Cascade To Niagara Converter plugin

Enabling this plugin will add a new option to the menu when you right-click on a Cascade particle system asset in the Content Browser.

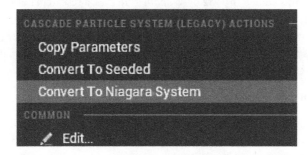

Figure 1.3: Converting a Cascade system into a Niagara System using the Converter plugin

This option will create a new Niagara System in the same folder as the Cascade system with the suffix **_Converted** added to it. The conversion does not fully support all cases, so expect to find a bunch of errors showing up in the Niagara System when you open it in the Niagara Editor, which will need to be manually fixed. So, while we can use the converter as a starting point to convert Cascade systems into Niagara Systems, additional work is almost invariably needed to complete the conversion.

To sum up, UE5 contains two different particle systems: Niagara, the primary particle system, and Cascade, which is available for compatibility purposes. In the next section, let us get an overview of the Cascade system.

Cascade particle system

Cascade is a modular particle effects editor integrated into Unreal Engine. As of Unreal Engine 5.1, the option to create a Cascade particle system has been moved into the **Miscellaneous** section.

Figure 1.4: Creating a new Cascade particle system in Unreal 5.1

Alternatively, if you upgrade from a UE4 project containing Cascade particle systems, you should be able to double-click on the particle system asset and open **Cascade Editor**. If you need to create a new particle system, you should use Niagara. This chapter is to familiarize you with Cascade just enough for you to manage any older projects that you might need to work on.

Learning about Cascade, Unreal Engine's previous particle system, can still be important for several reasons:

- *Legacy content*

Many older projects in Unreal Engine still use Cascade, so being able to work with it is an important skill to have in case you need to make changes or updates to existing content.

- *Historical context*

  Understanding Cascade can give you valuable insights into the history and evolution of Unreal Engine and how it has changed over time to become the powerful development platform it is today.

- *Skills transfer*

  Many of the concepts and techniques used in Cascade are still applicable to Niagara, so learning about Cascade can help you build a strong foundation of particle system knowledge that you can apply to your future work with Niagara.

- *Career opportunities*

  There may still be opportunities to work with Cascade in certain industries, such as film and television, where older projects may still be in use.

While Niagara has replaced Cascade as Unreal Engine's current particle system, learning about Cascade and its capabilities can still be a valuable part of your education as an Unreal Engine developer.

If you do not plan on working on any old UE4 projects, feel free to skip this chapter and move on to *Chapter 2*.

Let's take a look at how Cascade works before leaping into Niagara.

Cascade particle systems are available as a part of the Starter Content pack. Starter Content is a bunch of assets made available in Unreal Engine as a starting point for the user to have some basic assets to work on at the start of a project. It has a bunch of audio files, textures, materials, meshes, particle systems, and other assets that you might typically need to prototype a project.

So let's get started: create a new UE5 project and enable **Starter Content**.

You can enable Starter Content by checking the **Starter Content** checkbox when creating a new project, as shown in *Figure 1.5*.

Figure 1.5: Ensuring that the Starter Content checkbox is ticked

Once the project is created, you should see the **Starter Content** folder in the **Content** folder. In the **Starter Content** folder, open the **Particles** folder in which you will find a few sample Cascade particle systems.

Figure 1.6: The legacy Cascade particle systems

Double-click on the **P_Fire** system to launch the Cascade interface.

Figure 1.7: The Cascade interface

There are six main zones in the Cascade interface:

- **MenuBar**
- **ToolBar**
- **Viewport**
- **Emitters Panel**
- **Details Panel**
- **Curve Editor**

We won't go into the details of the **Cascade Editor**, but we will review some key features in the Editor that are relevant to our pursuit of learning about Niagara:

- The **MenuBar** section has standard **Save**, **Undo**, **Redo**, and more menu options
- The **Toolbar** section also has the **Save**, **Undo**, and **Redo** options along with a few notable buttons as follows:

  - **Restart Sim**: This resets the particle simulation in the **Viewport** window
  - **Restart Level**: This resets the particle system instance in the level
  - **Thumbnails**: This takes a snapshot of the **Viewport** and saves it as a thumbnail in the Content Browser
  - **Bounds**: This toggles the **Bounds** display in the **Viewport**
  - **Origin Axis**: This toggles the origin axis of the particle system in the **Viewport**
  - **Background Color**: This changes the **Viewport** background color
  - **Level of Detail (LOD)**: This opens the options to create and modify particle LODs

The **Viewport** shows a real-time preview of the particle as it would appear in game. It also has different render modes such as **Unlit**, **Wireframe**, and **Shader complexity**, which can be accessed via the View modes submenu. You can also play the system at different speeds such as *100%*, *50%*, *25%*, and *1%*. There are a lot of properties available in the View menu to visualize different aspects of the particle system, about which we won't go into detail.

The **Viewport** can be navigated using the **left mouse button** (**LMB**) to tumble the camera, **middle mouse button** (**MMB**) to pan the camera, and **right mouse button** (**RMB**) to rotate the camera, *Alt + LMB* to orbit the system, and *Alt + RMB* to dolly.

Emitters panel is the main work area of the Cascade particle editor. This is where you create all the emitters contained in the particle system. You can also add and modify different modules to the emitters. Modules control various the behavioral aspects of the particles released by the emitter. A module can interact with other modules and this interaction is affected by their position in the stack of modules. So, for example, if we have two modules applying different velocities, it will result in the cumulative velocity of those modules being applied to the particles.

Creating an emitter

Now that we've learned a little about the Cascade particle system, let us see how we can add a new emitter to the **P_Fire** particle system. A Cascade emitter can only be created inside a Cascade particle system.

Figure 1.8: Creating a new Particle Sprite Emitter by right-clicking in the blank space

You can add an emitter by right-clicking on the blank area in the panel and clicking on **New Particle Sprite Emitter**.

Figure 1.9: Newly created Particle Emitter with the emitter Block at the top and the modules below

The emitter created is a column with an **emitter block** on top and a few default modules under it. The emitter block contains the main properties of the emitter, which can be accessed by clicking on it. On clicking the emitter block, the **Details** panel shows properties including **Emitter Name**, **Emitter Render Mode**, and **Detail Mode Bitmask**, which can be edited. You can also change the color of the color bar on the left of the emitter block here by changing the **Emitter Editor Color** setting, allowing you to color-code your emitters:

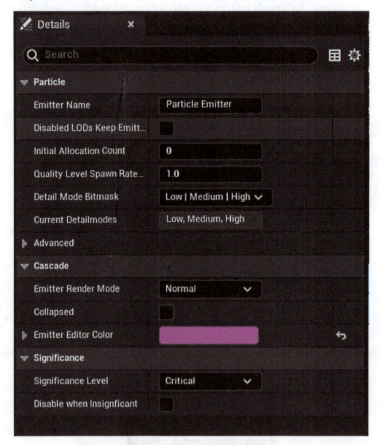

Figure 1.10: The Details panel in the Cascade Editor

In each emitter, we can add modules (which are components of the emitter) to modify particle behavior. A module can, for example, affect the velocity, direction, color, and other properties of a particle. Every emitter will have a **Required** module and a **Spawn** module.

The **Required Module** has all the must-have properties of an emitter. These properties include properties that describe the material applied to the particles, the position of the emitter origin, any rotation applied to the emitter, and the alignment of the particle with respect to the screen. Many of these properties will be covered in the upcoming chapters in the context of Niagara.

The **Spawn** Module contains the properties that affect the way particles are spawned. In this module, the **Spawn** and **Burst** categories determine the rate at which particles are spawned.

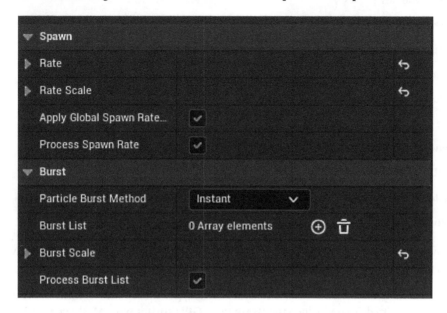

Figure 1.11: The Spawn Module properties in the Details panel

After the **Required** and **Spawn** modules, you can add any number of modules as required to get the effect that you want. The modules can be added by right-clicking on the emitter column.

These modules can be divided into the following categories depending on their function. The functions of the modules in each category should be evident from their names:

- **Acceleration** modules
- **Attractor** modules
- **Beam** modules
- **Camera** modules
- **Collision** modules
- **Color** modules
- **Event** modules
- **Kill** modules
- **Lifetime** modules
- **Particle** lights

- **Location** modules

- **Orbit** modules

- **Orientation** modules

- **Parameter** modules

- **Rotation** modules

- **Rotation Rate** modules

- **Size** modules

- **Spawn** modules

- **SubUV** modules

- **Vector Field** modules

- **Velocity** modules

As we learn more about Niagara, you will find that these Cascade modules have Niagara equivalents to allow us to recreate any of the effects produced in Cascade.

In addition to the aforementioned modules, we also have **TypeData modules,** which determine the type of particles emitted. These include the **Beam** type, **GPU sprites** type, **Mesh** type, and **Ribbon Data** type. The type of particles each of these emits should be evident from their names. As you would expect, Niagara has equivalent methods of its own to determine the type of particles emitted by the Niagara emitters.

Finally, we have the **Curve Editor**. This used to be the standard Unreal curved editor interface. This interface has since changed in other areas of Unreal Engine (including Niagara); however, it is somewhat frozen in time when it comes to Cascade. The Curve Editor allows the user to modify any values specified in any module that will change across the lifetime of a particle (or an emitter). To make a property editable in the Curve Editor, we need to set that property to **DistributionFloatConstantCurve** in the **Details** panel.

To push any module property to the **Curve Editor**, click on the green box on the left of the module.

Figure 1.12: Click the rightmost green box on a property to add it to the Curve Editor

To remove any curve from the **Curve Editor**, right-click on the property and click **Remove Curve**.

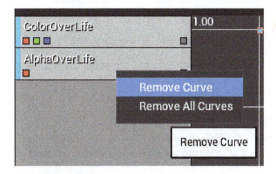

Figure 1.13: Remove Curve from Curve Editor

Curve Editor has tools to add/edit points to the curves, which are accessible from the toolbar at the top of **Curve Editor**.

Figure 1.14: The Curve Editor toolbar has tools similar to those in other animation apps

This briefly sums up the Cascade particle system's interface. Many of the concepts and tools used in Cascade will be discussed in detail when we learn about Niagara.

The reasons behind Niagara's development

With an expanding user base, the need for a particle system that was more robust and worked across all industries was felt more acutely. The use of Unreal has now extended beyond gaming to industries including architectural visualization, automotive and industrial design, virtual production, and training simulations. This had led to demands for accurate, efficient, and easy-to-use particle workflows. The newer workflows demanded that artists should be able to work on particle systems easily without having to deal with a complex set of tools while also letting the technical team members have access to tools that may not be very user-friendly but allow them to create customized solutions. Particle systems also needed to be more fully integrated with the main code of the Unreal app.

Against the background of these scenarios, the shortcomings of the Cascade particle system started becoming more evident.

The development of Niagara was driven by the following goals:

- It should be easy to use and put control in the hands of the artists
- It should be customizable and programmable in every aspect
- It should have an improved toolset for operations such as debugging, visualization, and performance
- It should be able to seamlessly interface with other parts of Unreal Engine

Perhaps the biggest issue with Cascade was that it was very difficult to add additional features or customized behaviors to particle systems. Artists were heavily dependent on programmers to add new features. Niagara has made it possible for artists to develop additional features on their own, giving them more control.

Every aspect of Niagara can be customized. Cascade does not offer such flexibility. In Niagara, every parameter of forces acting on particles can be tweaked and connected to external parameters. The user can, for example, drill down and change the force of gravity acting on an object over time using a sine wave. This open-ended architecture puts no limits on the kind of effects that can be achieved with Niagara.

Cascade used CPU resources very inefficiently. CPU and GPU simulations would work very differently. Niagara is optimized to handle both GPU and CPU sims and achieve parity between them. Niagara also has two great tools for debugging simulations. In the Niagara Editor, you can use the **Debug Drawing** tool to see visual representations of the particle system, while in the game level you can use **Niagara Debugger**, which shows detailed information about given particle systems in the heads-up display. This helps pinpoint performance and behavioral issues in your particle systems easily.

Niagara works very well with other parts of Unreal Engine. For example, a game object's speed data can be shared very easily with the Niagara particle system to drive various parameters in the particle system, such as sprite size, the brightness of the particle, or the amount of gravity acting on the particle. This lets game designers create fine-tuned game mechanics very easily in a short amount of time. You also read data from external sources.

Cascade did have some upsides. The module-stacking workflow was a great way to get an overview of the particle system at a glance, and Cascade was very approachable for non-technical artists. However, the node graph paradigm of Unreal is very powerful and was necessary to adopt to deliver the next-gen features promised by Niagara. So, a hybrid method with both a stack and graph was chosen for Niagara, which derives the advantages of both paradigms.

Figure 1.15 illustrates the stack paradigm in Niagara where modules are stacked on top of each other. The stack-based workflow is simpler and suitable for designing basic particle behaviors.

Figure 1.15: The stack paradigm in a Niagara Emitter node

While the stack paradigm is simpler, it can be a bit limiting in its flexibility and hence its capabilities. Therefore, when such flexibility and power is required, a node-based approach is used, as shown in *Figure 1.16*. You will find Niagara adopting a node-based workflow for designing Niagara modules.

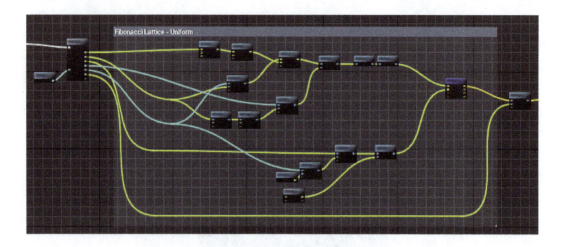

Figure 1.16: The graph paradigm inside a Niagara module

As we will learn later in the book, Niagara also makes it easy for teams to work in parallel developing particle systems by employing a modular approach to development and eliminating production bottlenecks.

All these reasons have helped Niagara replace and improve upon the old Cascade particle system and leave it perfectly poised to take on the challenges of delivering particle effects for the wide variety of industry verticals in which Unreal Engine 5 finds itself being used.

Use cases for Niagara

Being the next-generation FX system, Niagara allows technical artists to add custom functionality to a particle system. It is equally accessible to beginner and advanced users alike. Beginners can start with a variety of templates as their starting points, while advanced users can add custom modules to create complex effects.

Niagara can be used to create all the effects that Cascade can and then go much further. Standard particle effects such as fire, smoke, rain, and snow are surprisingly easy to set up.

Niagara particles can be used to create much more beyond standard particle effects. As you gain more knowledge of Niagara, you will find interesting and powerful features that extend its capabilities.

Some of the advanced features include interfacing with the world by reading mesh triangles, tracing against physics volumes, and reading scene depth and query distance fields.

These features allow you to create flocks of birds or swarms of spiders that respond to the game environment. A flock of bats, for example, can work their way through an enclosed cave environment without colliding with the rocks. A swarm of spiders made in Niagara can crawl across the floor, over any obstacles, and react to the presence of a player. Particles can be represented by animated meshes to render more authenticity to such a simulation.

Niagara also makes it possible to create complex effects such as the morphing of meshes with a particle-based transition where an object may dissolve into particles, and then those particles reassemble to form another object of a different shape. The objects, in this case, can be static or skeletal meshes, which can help game designers include interesting events in their games.

The most important aspect of all the aforementioned effects is you do not need a programmer to design these effects. Unreal artists can design such effects on their own without needing any programming knowledge.

Niagara makes it extremely easy for other parts of Unreal to share data with the particle system. This allows you to build blueprints that include Niagara Systems and have exposed variables with which even a beginner can tweak the Niagara System to produce an array of effect variations. For example, a **SnowStorm** blueprint can contain abstracted values such as snow density, snowflake size, and turbulence in the **Details Panel**, which will act on the appropriate parameters in the Niagara System (**Spawn rate**, **Sprite size**, **Curl force**, etc.) to create various stages of a snowstorm or even change the values in real time in a game in response to player actions. These types of blueprints allow junior Unreal artists to indirectly modify Niagara System properties without requiring extensive knowledge of Niagara.

A Niagara System can also read external data such as a sound file and have the particle systems react to that data to create a visualization effect. For example, we can have a particle system spawn different colored particles in response to different frequencies of a soundtrack driving the particle system. For a more visual representation of what is being discussed here, you can go to `https://www.youtube.com/watch?v=Vg1niqfDuzs`.

In UE5, Niagara goes much further with **Niagara Fluids**, a plugin that adds real-time grid-based simulations to Unreal Engine. This plugin adds templates for **two-dimensional gas simulations** (**2D Gas**), **three-dimensional gas simulations** (**3D Gas**), **Fluid Implicit Particle** (**FLIP**), and shallow water simulations, which are easy enough for beginners to use and can be tweaked at a granular level by advanced users.

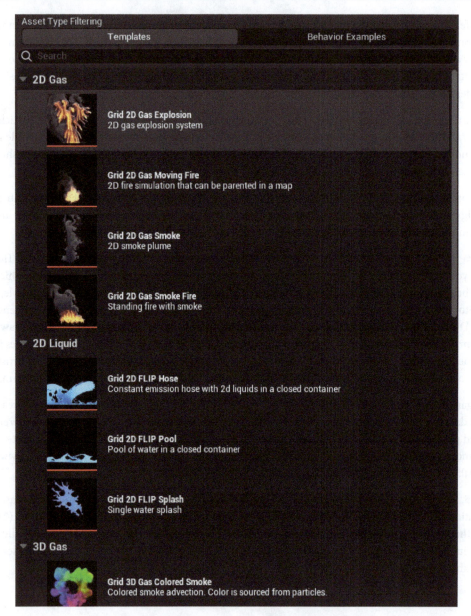

Figure 1.17: Niagara Fluids templates in Unreal Engine 5

While the fluid simulations can look amazing, not all devices on which your game may run will have enough resources to support a real-time fluid simulation. To bypass that, UE5 includes the **Baker** module to bake fluid simulations into a flipbook. These flipbooks can be used with sprite-based particle systems, which are very efficient and can run on slower devices. This also enables us to create rich secondary effects without sacrificing performance.

Niagara in UE5 also has a GPU ray-tracing option in the **Collision** module. Traditionally, GPU particles have used depth buffer to approximate collisions. Collisions using depth buffer are not very accurate. When using depth buffer, if the GPU particle system is occluded by an object, the particle system is culled by Unreal Engine and disappears. This can look odd. With the new hardware-based raytraced collisions, you can rest assured that the collisions will be accurate, and we will not encounter situations where GPU particles disappear.

In the following chapters, we will be building up an Unreal level from scratch and exploring a few of these use cases just discussed.

Summary

In this chapter, we briefly introduced the legacy Cascade particle system. This is important because, while the Cascade particle system is a legacy system, being familiar with Cascade will help you work on any old projects which use Cascade; not only that but Cascade particles are also in extensive use in many ongoing projects in many companies. There are also a lot of Cascade-based particle systems in the Unreal Engine Marketplace that you may need to incorporate in your project (be they current or future projects), so it is worth developing this ability now. After that, we explored the design paradigms in the new Niagara System and discussed the reasons behind the development of Niagara (despite Cascade still being functional). Finally, we looked at the new features expected to arrive in upcoming Niagara versions. In the next chapter, let us delve into some basic particle concepts to lay the foundations for our further learning.

2

Understanding Particle System Concepts

Particle systems are a technique used in computer graphics to simulate a variety of phenomena, such as fire, smoke, rain, and explosions. In a game engine, particle systems are used to add realism and visual interest to scenes. The three main functions of particle systems are emission, simulation, and rendering.

Emission is the process of creating and releasing particles from a specific location or object. This is controlled through properties such as emission rate, velocity, and direction. Emission can also be triggered by events, such as collisions or user input.

Simulation refers to the movement and behavior of the particles over time. This can include properties such as gravity, wind, and turbulence, which can be used to create realistic movement patterns. Particles can also be affected by other forces, such as collisions and user-defined behaviors.

Rendering is the process of displaying the particles on the screen. This can be done using sprites or 3D models and can include techniques such as billboard rendering and alpha blending to create realistic effects.

In this chapter, we will see how these apply to Niagara. We will also learn a few concepts of vector mathematics, which we will apply to our particle systems in the later chapters. Finally, we will learn about other particle system tools available on the market.

This chapter will cover the following topics:

- Exploring key particle concepts
- Vector mathematics and matrices and their representation in Niagara
- Understanding particle system tools

Technical requirements

You can find the project we worked on in this book here:
`https://github.com/PacktPublishing/Build-Stunning-Real-time-VFX-with-Unreal-Engine-5`

Exploring key particle concepts

In this chapter, we will understand the key functional concepts of particle systems and how they are implemented in the particle systems in Unreal. All particle systems will invariably have some variation of the following functionalities built into them:

- Emission
- Simulation
- Rendering

Each one of these functionalities will have some variables that will be exposed to enable the user to tweak the behavior of the particle system. Most software has this functionality built in a monolithic element with limited ability to customize it. In Unreal, it is built into modules that are stacked on top of each other. The modules have exposed variables, which can be used to tweak the module's effect on the behavior of the particle system. These modules are roughly grouped into three broad functionality groups: Emission, Simulation, and Rendering.

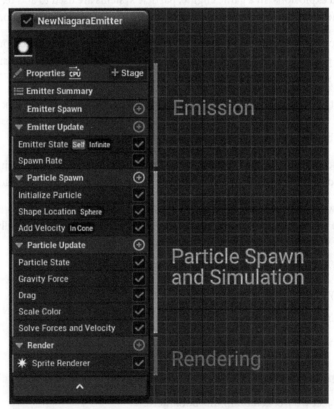

Figure 2.1: The Niagara Emitter node grouped into three functionalities

So, let's take a look at these functionalities.

Functionality

Now, we will delve deeper into the individual functionalities – emission, simulation, and rendering.

Emission

This part of the system handles the spawning of the emitters, as well as their behavior throughout their lifetime. In Niagara, emitters and systems exist as independent entities. Emitters need not always be a part of a particle system. The emitters can be called into any particle system as a reference. This structure allows us to work on the emitters and systems independently, making the workflow modular and scalable.

Multiple emitters can be called in a system to create the required effects. For example, for a fire system, we may have an emitter for flames, another emitter for embers, and another one for smoke. All three emitters will come together in a fire system to create the overall effect of fire. Emitters can also be affected by the behavior of other emitters in the system using event handlers, which allow users to create event-based effects.

Emitters determine how particles are emitted – for example, from a point space, volume, the surface of a geometry, or from the surface of an animated skeletal mesh. This is how the emitter section looks in Niagara:

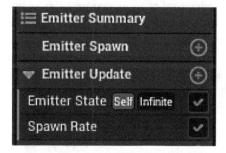

Figure 2.2: Emission-related functionality grouped together

Simulation

This is where the majority of a particle system's behavior is determined. Once the particles have been emitted, the simulation *modules* take over and determine things such as the movement, the change in scale or color, or the animation of sprites in the particle system. Each module will affect a specific aspect of the particle system's behavior. These modules use mathematical formulas, randomized values, and cached data to affect the behavior of the particles. The modules are usually stacked, and the order of stacking affects the final outcome of the particle behavior. The efficiency of the simulation modules matters and various methods such as using GPU acceleration are used to enable particle systems to handle a large number of stimulated particles. This is what a typical simulation section looks like in a **Niagara Emitter** node:

Figure 2.3: Simulation-related functionality grouped under Particle Update

Rendering

While all the simulation calculations are done by the modules that handle simulation, the particles won't be visible unless they are rendered by the particle system. This is handled by the rendering module. The fact that the rendering module is independent of the simulation module makes it possible to render the simulated particles in different forms such as sprites, meshes, or ribbons. The same simulation can create different images based on the rendering module chosen. For example, the rendering module, depending on its type, could render a set of simulated particles as two-dimensional sprites, as a three-dimensional geometry mesh, or maybe as ribbons. The simulation remains the same; only the rendering module changes. You can also choose to simultaneously use multiple rendering modules in the same particle simulation. This is how the rendering section looks in Niagara:

Figure 2.4: Rendering-related functionality grouped together

In Niagara, the three functionality groups discussed here manifest themselves as *stages*. These stages have names ending with **Update** and **Spawn**; for example, **Particle Update** and **Particle Spawn**. We also have advanced stages such as **Event**, or **Simulation**, about which we will learn in the later chapters.

Next, let's talk about module groups.

Module groups

Unreal takes a modular approach where the particle system starts out as a simple entity with a few essential properties. These properties and the additional properties required to add functionality to the particle system are made available as modules. Users can add and remove these modules from their particle system as required to achieve the desired behavioral effect. Each module will influence a specific aspect of the particle behavior. Certain modules can alter the color of particles throughout their lifetime, for example, transitioning from being yellow at birth to red at death. Other modules simulate forces such as wind and gravity on the particle system. Since the user only adds the modules that they need, the particle system is lean and efficient.

The particle simulation data flows from top to bottom of a Niagara module stack. These modules can be assigned to a group. The group determines where the module will be executed. A module in a Particle group, for example, would be executed on particles only and not on emitters. The groups are associated with **namespaces** that determine what data the module can affect.

There are three module groups which are as follows:

- System
- Emitter
- Particle

Niagara has many namespaces. Some of the important ones are **System**, **Emitter**, **Particle**, **Engine**, and **User**.

Namespaces may seem a bit confusing for a beginner, so think of them as an additional qualifier to help segregate similarly named variables.

For example, **SYSTEM LoopCount** is different from **EMITTER LoopCount** due to them being in separate namespaces, that is, **SYSTEM** and **EMITTER**.

Figure 2.5: Namespaces in Unreal help differentiate similarly named
parameters according to their mode of operation

The namespaces restrict the places from where the data can be read or written to. For example, the **Engine** namespace parameters are read-only values that come directly from Unreal Engine itself. The **Engine** namespace parameters enable us to access information such as **Delta Time** or the **Position** and **Velocity** settings of the owning actor, which are properties handled by the game engine. **User** namespace parameters also exist, which can be created by a user for arbitrary use.

In *Figure 2.6*, we see an example of a parameter called **PacktProperty** created for an arbitrary use case, which will have to be defined elsewhere in the system.

Figure 2.6: A custom property PacktProperty created in the USER namespace

We will go much deeper into the Niagara implementation of these concepts in upcoming chapters. This chapter should give you a broader idea of the concepts used in particle systems and how those concepts would translate to a Niagara workflow.

One of the most important mathematical concepts that you need to be familiar with when working with particles is the concept of vectors and matrices. While we do not use matrix transformations in this book, we do use vectors a lot. Let us refresh our understanding of vector mathematics and matrices before we jump into Niagara. Jumping into Niagara Falls is not recommended.

Vector mathematics and matrices and their representation in Niagara

Vector mathematics and matrices are one of those boring subjects in high school mathematics that you may have entirely skipped because you had no idea where you would use them in your daily life. Well, now you know. They are used everywhere in games and particle systems in particular. Let us take some time to brush up on our concepts, starting with vector mathematics.

What is a vector?

Let us ignore all the complicated talk about magnitude and velocity used in the standard definition of a vector and have a look at the bare essentials.

In a two-dimensional space, a vector just holds an x and y value; for example, vector a = (3, 4) where 3 is x and 4 is y.

This representation can be used to define the location or direction of an object. We can also find the speed of an object by finding what is known as the magnitude of the vector. Let us see how a vector is used to define a location and then we will see how it can also be used to define a direction.

A vector as a location.(coordinate)

Let's look at vector as a location.

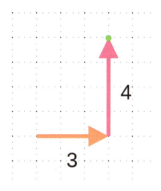

Figure 2.7: Representation of a 2-dimensional vector (3,4)

In the preceding diagram, the vector a represents the position of a point in 2D space at 3 units in the *x* direction and 4 units in the *y* direction. In 3D space, we can define the location of a point using a Vector3d as shown here:

Vector3d a = (3, 4, 5)

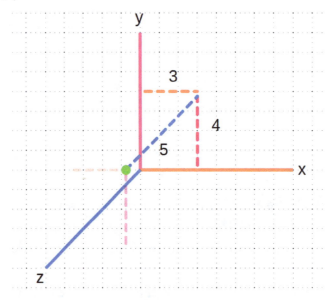

Figure 2.8: Representation of a three-dimensional vector (3,4,5)

In this diagram, we have added an additional axis, the *z* axis, which determines the height of the point from the *xy* plane, thus defining the point in 3D space.

Now, let's see how the same vector can also be interpreted to define a direction.

A vector as a direction

For vector a = (3, 5), we can find the direction and magnitude of the vector by breaking it into components.

This is how we can do it:

1. Let us move in 3 units in the *x* direction (denoted by segment AB).
2. Then from point B, move 5 units in the *y* direction (denoted by segment BC).
3. We can now draw a segment AC, the orientation of which represents the direction of the vector.

Here, the length of the vector gives us the magnitude of the vector.

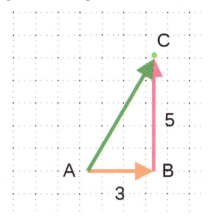

Figure 2.9: Segment AC representing the direction of a vector of value (3,5)

This method of determining the direction using a vector can be extended to 3D space by having a 3D vector.

In Niagara, a 2D vector is called **Vector 2D** while a 3D vector is just called **Vector**. This is because a 3D vector is the most commonly used vector. A 4D vector also exists, which is called **Vector 4**. **Vector 4** is generally used to define color in Niagara, with the first three values of the vector representing the RGB values of the color and the fourth value representing the alpha channel value. 2D vectors are used to define 2D data such as UV coordinates.

The following figure shows how they appear when defined in Niagara:

Figure 2.10: A 3D, 2D, and 4D vector as represented in Niagara

Now that we understand vectors, let us look at vector operations. Vector operations can help us work with multiple vectors. For example, if two vectors act on an object, we can find the resultant vector by adding those two vectors together.

Vector operations

In particle systems, the movement behavior of particles is a result of a number of vector forces. The calculation of the resultant force acting on a particle is done in the modules. We use various vector operations to calculate these resultant forces.

The resultant vector force acting on a particle can be calculated using vector operations such as addition, subtraction, and multiplication. Let us understand how these vector operations work and see how they are implemented in Niagara.

Addition

Let us see what it means to add two vectors a = (3,4) and b = (5, 2) together.

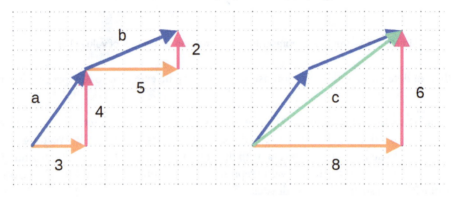

Figure 2.11: Vector addition

Consider the diagram here. We break both the vectors, **a** and **b**, into components x and y. Then, we add the x components and the y components.

In our case, let's say the sum of our vectors **a** and **b** is a new vector, **c**. So:

$c = a+b$

$c = (3,4) + (5,2)$

$c = (3+5, 4+2)$

$c = (8,6)$

We see that resultant vector $c = (8,6)$, when calculated, corresponds to vector **c** shown in the diagram, which is the resultant vector for the two vectors **a** and **b**.

Subtraction

We can also subtract vectors.

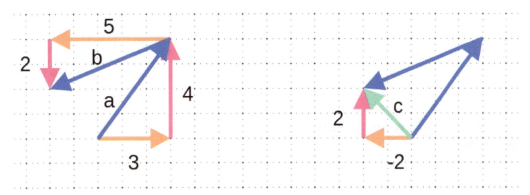

Figure 2.12: Vector subtraction

Let's subtract vector **b** from vector **a**. The resultant vector, **c**, can then be denoted as follows:

$c = a - b$

$c = a + -b$

$c = (3,4) + -(5,2)$

$c = (3-5, 4-2)$

$c = (-2,2)$

Like with the addition of vectors, in this case, we see that the resultant vector $c = (-2,2)$ corresponds to vector **c** in the diagram

In this explanation of Vector operations, we considered 2D vectors for our ease of understanding, but usually in Niagara, as we deal with 3D space in our games, you will find yourself using 3D vectors.

Magnitude and unit vectors

We know that the length of a vector is equal to its magnitude.

To make calculations easier, we sometimes consider the magnitude of certain vectors to be equal to 1. These vectors are called unit vectors. Unit vectors are also called direction vectors, as they help convey direction – their magnitude does not matter, as it is always equal to 1.

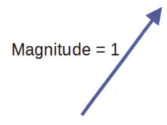

Figure 2.13: A unit vector

You can multiply a unit vector by a scalar to increase the magnitude of the vector in the vector's original direction.

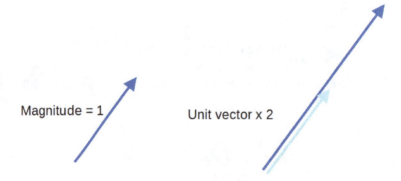

Figure 2.14: A unit vector multiplied by a scalar of value 2

We now understand the concept of vector operations and their representation in mathematics. Let us see how these operations are represented in Niagara.

Representation of vector operations in Niagara

These vector operations are used a lot in Niagara. Thankfully, they are handled by nodes, so you do not have to actually do the calculations yourself. The following are some of the nodes that we find in Niagara for addition, subtraction, and multiplication:

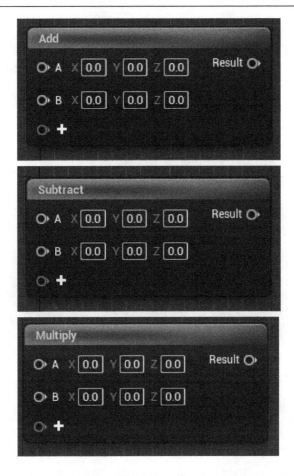

Figure 2.15: Addition, subtraction, and multiplication of vectors implemented as nodes in Niagara

We can also multiply a scale value (or simply put, a float or an integer value) to an integer similar to how you multiply a unit vector.

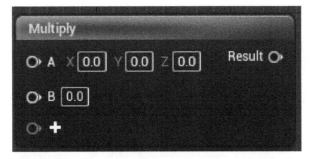

Figure 2.16: Multiplying a vector with a scalar as represented in
a Niagara node (note that the result is a vector)

If you are wondering where you can find these nodes in Niagara or are unable to find them, please be aware that in their original form, they look like the following figure, and the input and output pins need to be converted into the appropriate data type as shown by right-clicking on the Numeric data type pin.

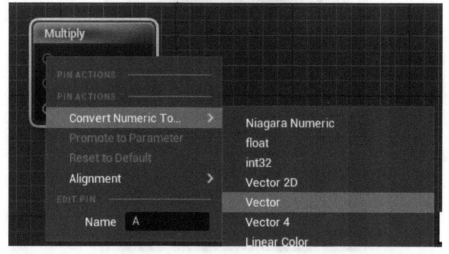

Figure 2.17: Converting a Niagara numeric data type pin into a vector pin

Take some time to explore the **Convert Numeric To...** option and see how the node with blue Niagara numeric pins changes when converted into Vector 2D, Vector 4, and other data types.

Matrix operations

In addition to vector operations, we also find a lot of use for matrices in Niagara. Some of the essential matrix operations are shown here. Matrix transform operations are necessary if you want to change the direction or position of an existing direction vector or position vector. The mathematics behind this is a bit complicated, but you can get away with just understanding what it does to your particle system.

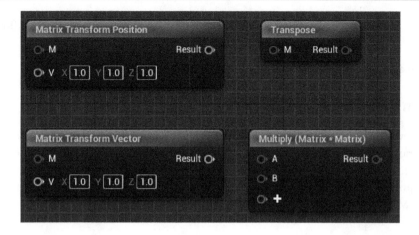

Figure 2.18: Matrix operations as represented by nodes in Niagara

A matrix transpose operation just flips the matrix across its diagonal. This operation is needed to modify particle properties.

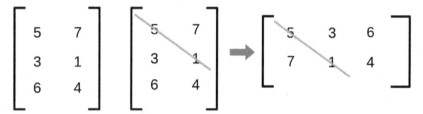

Figure 2.19: The matrix transpose operation

The preceding operation shows a matrix transposed across its diagonal.

After gaining familiarity with this basic vector, matrix operations, and their Niagara equivalents, we are ready to finally begin our journey to understanding this in Niagara.

In the next section, we will have a look at some existing particle system tools available independently or as a part of other software that are commonly used in the industry.

Understanding particle system tools

There are a bunch of tools and off-the-shelf solutions available for creating particle effects. While you need not know all of them, you certainly will encounter some of them in your work as a real-time FX artist. Many studios might do part of a simulation in one of their own custom tools or one of the tools mentioned here. You may need to receive data such as vector fields from one of these tools and incorporate it into your Niagara System. You may also receive references for effects produced by one of these tools and may be required to recreate them in Niagara.

A lot of these tools have been around for quite some time and have shaped the workflows and terminology used by FX teams. Niagara builds on these concepts.

The following are some of the most widely used particle system tools.

Afterburn

Afterburn, developed by *Sitni Sati*, is a volumetric particle effects plugin. It has been used in VFX for clouds, smoke, explosions, and other effects. It has been largely replaced by Sitni Sati's other fluid dynamics plugin, which is called **FumeFX**. Like most modern fluid simulation plugins, FumeFX offers instant viewport feedback. FumeFX supports 3ds Max, Maya, and Cinema 4D while Afterburner supports 3ds Max.

RealFlow

RealFlow, available as a standalone program as well as a plugin for **3ds Max**, **Maya**, and **Cinema 4D**, is a particle-based fluid and dynamics simulation tool used to simulate fluids, fluid interactions, and a wide variety of phenomena such as elasticity, granular particle flow, and viscous fluids. Developed by Next Limit Technologies, this is one of the leading simulation software used in the VFX industry.

Trapcode Particular

Trapcode Particular is a very popular plugin for After Effects, part of *Maxon's Red Giant suite* of tools. As with the other particle solutions listed here, Trapcode also uses particle emitters to create a wide variety of particle effects such as smoke, fire, and clouds. It has solutions to create flocking and swarming behaviors, as well as the capability to do physics and fluid simulation. This one also finds a lot of use in motion graphics design.

Particle Illusion

Particle Illusion is another excellent standalone particle simulation app developed by Boris FX. This application finds more use in motion graphics and titles and is used widely for audio visualization with particle effects due to a very easy-to-use feature it has called **Beat Reactor**.

nParticles

nParticles has existed in **Maya** for quite some time. It uses the Maya Nucleus dynamic simulation framework and is used to create fire, smoke, liquid, and mesh effects. nParticles also works seamlessly with other systems in Maya such as **nCloth** and **nHair**.

PopcornFX

PopcornFX is a very popular real-time particle FX solution. It supports multiple game engines, including Unreal Engine, and has been used in many leading games such as Forza Horizon 5, Age of Empires III, and Warcraft III to name a few. It comes with a PopcornFX editor, which helps create and manage assets related to particles in game engines. The PopcornFX workflow in many ways is similar to the basic Niagara workflow, having similar internal editors such as a timeline and curve editors. If you have experience with PopcornFX, you will quickly be comfortable in Niagara, and vice versa.

Houdini

Houdini, developed by SideFX software, has what are called **particle operators** (**POPs**) for creating dynamic particle effects. Houdini has tools to create complex particle systems and simulations. Houdini interfaces with Unreal Engine to make the power of Houdini available in Unreal Engine through Houdini Engine. With Houdini Engine, not only particle systems but also the full power of Houdini's procedural asset-building workflow can be accessed in Unreal Engine. Extensive use of the Houdini toolset was made in the Matrix Awakens demo during the Unreal Engine 5 release.

In conclusion, as a real-time FX artist, you will encounter a variety of particle system tools in your work. Each tool has its own set of features and capabilities, and it is important to understand the strengths and weaknesses of each tool in order to choose the right one for the task at hand. Additionally, you may need to receive data from one of these tools and incorporate it into your Niagara System or recreate effects produced by one of these tools in Niagara. Understanding these tools and how they work will help you become a more effective and efficient real-time FX artist.

With this, we come to the end of *Chapter 2*.

Summary

In this chapter, we familiarized ourselves with the groups into which Niagara modules can be classified. We learned about namespaces, a feature we will use a lot, and we brushed up on some high-school mathematics about vectors and matrices. We also learned about other particle system tools used in the industry. Let us now move on to the next chapter, where we will come to understand the Niagara architecture in detail.

3

Exploring Niagara Concepts and Architecture

Unreal Engine's Niagara is a robust and versatile system for creating real-time visual effects. With its hybrid stack and node paradigm, it provides an intuitive and modular workflow for creating and manipulating particles, vector fields, and custom logic. The ability to expose variables to other parts of the engine makes it easy to integrate Niagara effects into your project, while stack groups allow for the organization and management of multiple effects within a single project. Whether you're creating simple particle systems or complex fluid simulations, Niagara's powerful features and flexibility make it a go-to choice for VFX artists and game developers alike.

In this chapter, we will cover the following:

- The Niagara architecture

Technical requirements

You can find the project we worked on in this book here: https://github.com/PacktPublishing/Build-Stunning-Real-time-VFX-with-Unreal-Engine-5

The Niagara architecture

As we saw in *Chapter 1*, the **Cascade** particle system started to show its age as the demands of the different industries using Unreal began to grow. Niagara was created to meet these demands while at the same time being powerful and easy to use.

Niagara manages to do this by incorporating the following features in its design architecture:

- All data is exposed to the user. Niagara can use data from any part of Unreal Engine as well as from other applications. This lets the user create interesting relationships between the particle system and other aspects of the game.

- Exposing all data can be overwhelming for the user. To make it easy for the user to work with the data, this data is classified hierarchically. This is done by using namespaces.

Figure 3.1: Parameters under the PARTICLES namespace

For example, in *Figure 3.1*, the **PARTICLES** namespace carries all the particle attributes, while in *Figure 3.2*, the **EMITTER** namespace carries all the Emitter attributes. The **Age** parameter is named similarly in **Particle Attributes** and **Emitter attributes**, but due to them being in different namespaces, there will be no name clashes. The namespace therefore allows similarly named attributes to be classified correctly.

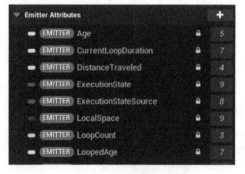

Figure 3.2: Parameters under the Emitter namespace

- The parameters can be of different data types, such as float, integer, linear color, Boolean, and so on. The data type is indicated by the color of the pill-shaped icon next to the parameter name. The colors are standardized throughout Unreal Engine. For example, the dark red color represents Boolean data, and the light green represents float data. These parameters can be accessed or written to in Niagara to drive the behavior of particle systems. Niagara combines the easiness of the stack paradigm with the power of the graph paradigm to create a hybrid workflow, giving the user the best of both worlds. In the Saack paradigm, the elements of the system are organized by stacking them on top of each other and the flow of operations is from the top of the stack to the bottom. In a graph paradigm, they are in the form of nodes connected to form a graph, and the flow of data is along the connector lines of the graph. The graph paradigm is considered more powerful, as it allows for complex customization, while the stack paradigm is less complex due to its unidirectional flow.

The emitter and system nodes in Niagara follow the stack paradigm, as seen in the following figure:

Figure 3.3: The Niagara nodes use the stack paradigm with modules arranged in a stack

The modules follow the graph-based paradigm for their workflow, as seen in the following figure:

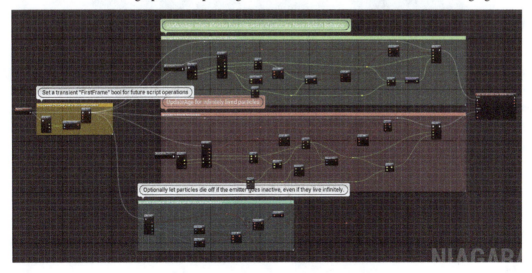

Figure 3.4: The Niagara modules consist of scripts developed using the graph paradigm

Next, we will take a deep dive into the hybrid structure.

Niagara's hierarchical hybrid structure

Let us understand how this hybrid hierarchy is represented as assets in Unreal and the relationship between these assets.

The hybrid workflow consists of three important components:

- Modules
- Emitters
- Systems

Figure 3.5 shows how these components show up in the **Content Browser** as assets.

Figure 3.5: An emitter, a module, and a system asset in the Content Browser

Figure 3.6 is a schematic of the hierarchy of these components and shows their relationship with each other, with the **Module** asset being at the bottom of the hierarchy and the **System** asset at the top, with each **Module** being a part of the **Emitter**, which, in turn, is part of the particle system.

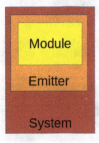

Figure 3.6: The hierarchy of module, emitter, and system

Now, let's take an in-depth look at the three components:

- **Module**: This is at the bottom of the hierarchy. Modules can be created using a visual **Node** graph using underlying **High-Level Shading Language** (**HLSL**) code. HLSL is the C-like high-level shader language that you use with programmable shaders and is beyond the scope of this book.

    This graph paradigm makes it very powerful. Modules can talk with other parts of Unreal Engine to generate complex behaviors. These modules can be stacked in emitters, which is the next level in this hierarchy.

- **Emitter**: Emitters use the stack paradigm. Emitters contain a stack of modules. The exposed parameters of the modules may be tweaked in the **Selection** panel, which appears in the Niagara editor and shows the properties of the selected module. The order of stacking is important, as a different order of stacking may lead to different effects. Emitters also have a timeline embedded into them, which helps control their temporal behavior.

- **System**: Systems also use the stack paradigm and are one step above emitters in the hierarchy. A system can contain one or more emitters. Like an emitter, a system has an embedded timeline. Like in an emitter, we can also modify the exposed parameters of a module in a system. The system can override any emitter or module parameter.

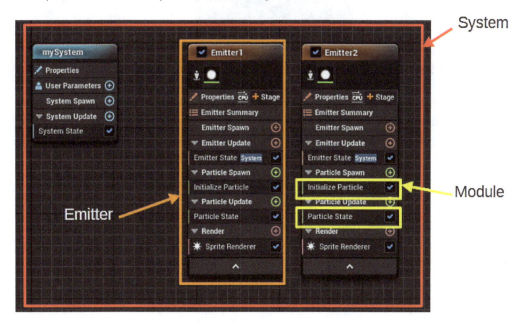

Figure 3.7: The module, emitter, and system as they appear in Niagara

Figure 3.7 shows the hierarchy elements as they appear in the editor.

Stack groups

As we just learned, modules are stacked in an emitter. The simulation flows from the top to the bottom of the stack. In order to make it easy to organize them, each module is assigned to a group. This helps determine where the module is executed. In *Figure 3.8*, the **Initialize Particle** module has been assigned to the **Particle Spawn** group, along with **Shape Location** and the **Add Velocity** module. This makes it clear that these modules act only at the moment that the particles spawn. The **Gravity Force** module, on the other hand, acts throughout the life of the particle, as it is under the **Particle Update** group.

Figure 3.8: Stack groups in an Emitter node (Particle Spawn, Particle Update, and Render)

As we develop particle systems in Niagara, we will see how the hierarchy and stack groups help shape the workflow. In the next chapter, we will create an emitter, a system, and finally a module and thus develop a full-fledged Niagara particle effect. As we progress through the book, you will become very familiar with these components, as we will use them regularly throughout the book.

Summary

In this chapter, we learned about the different features of the Niagara architecture. We learned about namespaces, the stack and graph paradigm used in Niagara, the different stack groups, and the hierarchy of the components of Niagara.

Understanding the different types of modules, emitter, and system components in the Niagara framework can help users better design and organize their Niagara Systems.

Modules, emitters, and systems are the building blocks of Niagara Systems, and by understanding the different types of modules available, users can select the appropriate components for the tasks they need to carry out.

4

Building Our First Niagara System

In this chapter, we will dive into the world of Niagara. We will begin by taking a tour of the Niagara Editor UI and exploring the various features it has to offer. After that, we will delve into the Niagara Emitter and learn how to create one from scratch. From there, we will move on to creating a Niagara System and adding it to a level and a Blueprint Actor. By the end of this chapter, you will have a solid understanding of creating a Niagara System from scratch.

In this chapter, we will cover the following topics:

- Exploring the Niagara Editor UI
- The Niagara Emitter
- Creating an Emitter
- Creating a Niagara System
- Adding a Niagara System to a level
- Adding a Niagara System to a Blueprint Actor

Technical requirements

Unreal Engine 5.1 or above is required. The installation procedure was explained in *Chapter 1*.

You can find the project we worked on in this book here:

```
https://github.com/PacktPublishing/Build-Stunning-Real-time-VFX-
with-Unreal-Engine-5
```

Exploring the Niagara Editor UI

To be able to learn Niagara, you need to be familiar with Unreal Engine 5 and have a basic to intermediate understanding of any 3D software and image editing software such as Photoshop or Affinity Photo.

In this chapter, we will get an introduction to the Niagara UI and create our first Niagara System.

Niagara is built as a plugin and is enabled by default. However, if you do not see the Niagara System option, check the Plugins panel by going to **Edit | Plugins** and ensure that the **Niagara** plugin is enabled, as shown in the following screenshot. The screenshot also shows the **NiagaraFluids** plugin, which adds fantastic real-time fluid features to Niagara.

> **Note**
>
> The NiagaraFluids plugin is still in beta mode, so exercise caution if you plan to use it in production.

Figure 4.1: Enabling the Niagara plugin

The method to create a Niagara asset has changed in Unreal Engine 5. In Unreal Engine 4 and below, we had to right-click in the Content Browser and then on **Fx | Niagara System** to create a Niagara System. In Unreal Engine 5, there is a newer, faster way, where you can right-click in the Content Browser and then on **Niagara System**. The **Particle System** option has been replaced by **Niagara System** in Unreal Engine 5.

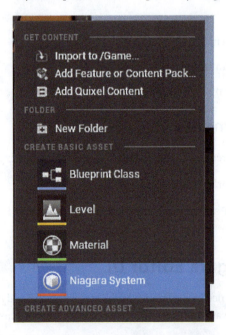

Figure 4.2: Creating a new Niagara System in the Content Browser

While the right-click menu encourages you to directly create a Niagara System, we are going to start with a different approach to creating a Niagara System. This will help us understand the modular functionality better.

To take full advantage of the modular functionality provided by the emitters in Niagara, we will start by creating a Niagara Emitter first, and then we will add this Niagara Emitter to the Niagara System.

Let's start by creating a Niagara Emitter:

1. To create the Niagara Emitter, right-click in the Content Browser and choose **FX**.
2. Here, you will find a bunch of options. From this, you can choose the **Niagara Emitter** option, as shown in *Figure 4.3*.

Figure 4.3: Creating a Niagara Emitter

3.   Clicking on **Niagara Emitter** will open the Emitter wizard.

4.   Let's start by choosing the **New emitter** option. This will create a new emitter from existing templates.

5.   Click on the **Next** button to move to the next screen.

Figure 4.4: The Emitter wizard

6.   The next screen will provide you with a list of templates. Let us choose the **Fountain** template. The particle system created using this template creates a simple fountain spray. Press the **Finish** button.

This will complete the wizard and create a Niagara Emitter asset in your Content Browser. You can rename the emitter to whatever you want. In this case, we are keeping the name of the emitter as **NewNiagaraEmitter**.

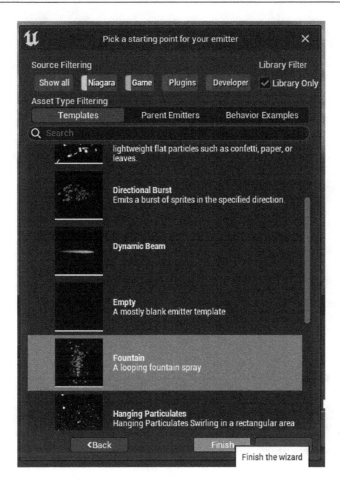

Figure 4.5: Choosing the Fountain template in the Emitter wizard

A Niagara Emitter has a light orange strip at the bottom of its icon.

Figure 4.6: The Niagara Emitter asset icon

We have created our first Niagara asset, which is a Niagara Emitter. We will continue working further on the emitter and studying it in detail. But before that, let us familiarize ourselves with the Niagara Editor UI. Once we know our way around the editor, we will come back to the emitter and start customizing it.

The Niagara Editor

As with every asset type in Unreal, we need to double-click on the asset in the Content Browser to open the appropriate editor. This opens the Niagara Editor for that emitter, as shown in *Figure 4.7*.

Let us understand the various parts of the editor.

There are ten key areas in this editor:

- **Toolbar**
- The **Preview** panel
- The **Parameters** panel
- The **System Overview** panel
- The **Local Modules** panel
- The **Selection** panel
- The **Timeline** panel
- The **Curves** panel
- The **User Parameters** panel
- The **Niagara Log** panel

We will mostly be working with the **Toolbar**, the **Preview** panel, the **System Overview** panel, the **Selection** panel, and the **Timeline** panel at the beginning and use the other panels as we go into advanced topics throughout the book. Let us see a few of these panels in detail.

Figure 4.7: The Niagara Editor

The Preview panel

The **Preview** panel shows us a preview of the emitter that we are working on. It enables us to get immediate feedback on various modifications that we do in the **Overview** node on the **System Overview** panel. The **Preview** panel has a hamburger menu on its top-left corner where you can toggle the **Realtime** option to enable real-time preview. Most of the time, you would want the **Realtime** option to be on. If it is off, you will see an additional status at the top of the viewport indicating that Realtime is off.

We can use the left mouse button to rotate around the previewed emitter if we are in **Orbit Mode** and look around in the scene if **Orbit Mode** is turned off. It is recommended not to turn off **Orbit Mode** as it can be difficult to get the particle emitter back into position if you move the camera around.

Figure 4.8: The Preview panel

We can use the right mouse button to zoom and the middle mouse button to pan. It is possible to drift away from the system when panning. We can also use the middle wheel to zoom in and out. We can press the *F* key to show all the particles, in case we have drifted away from the system or zoomed too close or too far away from the particle system. It is better to leave the rest of the options in the menu alone as they offer no significant advantages to our workflow.

We can use the **Show** menu to show options that help us diagnose issues and keep the emitter efficient. This includes **InstructionCounts**, **ParticleCounts**, and **Emitter Execution Order**.

Figure 4.9: The Show menu in the Preview panel

The Parameters panel

The **Parameters** panel shows us the available parameters with their name, type, status, and the number of times they have been called in the emitter. We can add custom parameters to an emitter by clicking on the + sign and choosing an appropriate parameter type that we need to add. We will come back to this panel when we do some advanced examples.

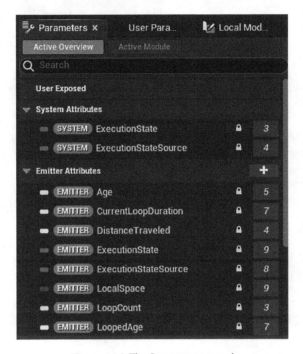

Figure 4.10: The Parameters panel

The System Overview panel

The **System Overview** panel shows the overview node for that emitter. We can use the right mouse drag to pan around in the viewport, the middle wheel, or *Alt* + right mouse drag to zoom in and out. These navigation functions may seem unnecessary when we are working on the emitter as we have just one overview node, but they do become useful when we are working on the Niagara System where we can have multiple overview nodes.

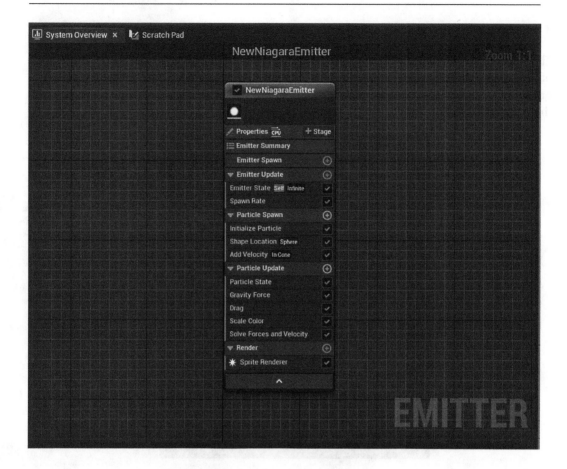

Figure 4.11: The Overview node

The Overview node is a stack of modules that define the properties and behavior of the emitter. We can add, remove, and disable modules in this overview node. To add a new module to the stack, we press the + sign on the right of the stack group label. This opens up a panel that lists all the modules available by default in the library. We will see at a later point in the book how to create our own custom modules. In this chapter, we will add some existing modules to our emitter to modify its behavior.

In Unreal Engine 5, we can now collapse the Overview node to reduce clutter. The modules have a color bar running along their left side to indicate the stack group that they belong to. For example, the **Emitter Update** modules have a red bar running along their left side. The **Particle Spawn** and **Particle Update** stack groups have a green bar running along their left side. We will be spending a lot of time getting familiar with this. Unreal also adds a helpful **EMITTER** label in a large font at the bottom right of the **System Overview** panel to prevent any confusion as to what part of the Niagara particle system we are working on.

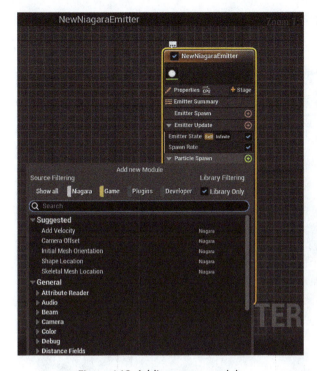

Figure 4.12: Adding a new module

The Local Modules panel

Next to the **System Overview** tab, we have the **Local Modules** tab. This is used to create and test out custom modules to be added to our stack. This is a bit of an advanced workflow, and we will visit it in a later part of the book.

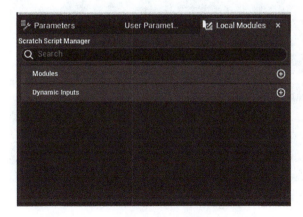

Figure 4.13: The Local Modules panel

The Selection panel

We now come to the **Selection** panel. Here, we see the details of the **Overview** node.

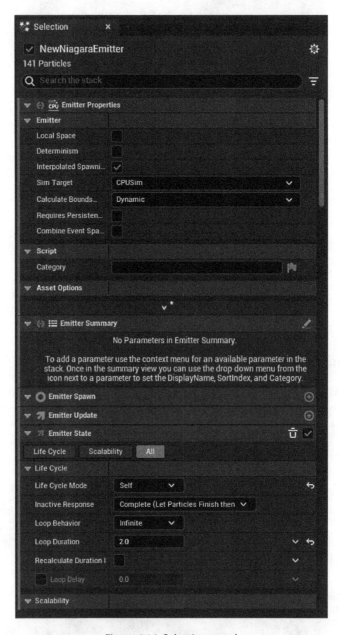

Figure 4.14: Selection panel

The amount of information on this panel may seem overwhelming at first, but we can narrow it down to the information we are interested in by just selecting the appropriate part in the **Overview** node. For example, if we click on the **Overview** node label, we see all the information associated with the node, which is a bit too much to handle at the beginning. We can focus on, say, the **Particle Spawn** module's information by clicking on the **Particle Spawn** label. This will only show the detailed information of the **Particle Spawn** group of modules.

Figure 4.15: Selection panel with only the Particle Spawn modules displayed

We can further narrow it down to just see the **Shape Location** module by clicking on the module in the **Overview** node. This will filter out only the information associated with the **Shape Location** module.

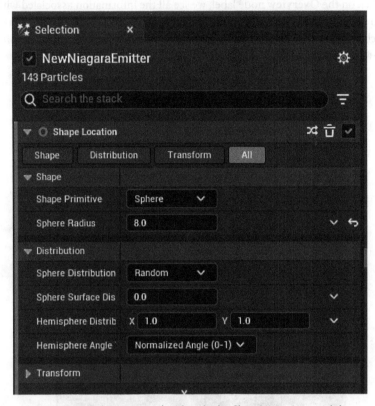

Figure 4.16: Selection panel with only the Shape Location module

Once we are familiar with this workflow, we can quickly and efficiently reach and modify the required parameters in the modules.

The **Selection** panel also allows us to enable/disable modules using the checkbox on the right of the detailed panel. We can also remove the module entirely by clicking on the trash can icon.

The **Selection** panel is also a place to add Dynamic Inputs of expressions to the parameters of the modules. This can be done by clicking the downward-pointing arrow on the right side of the parameter. In the following screenshot, we could let **Sphere Radius** be defined by a random value by choosing **Random Range** in float.

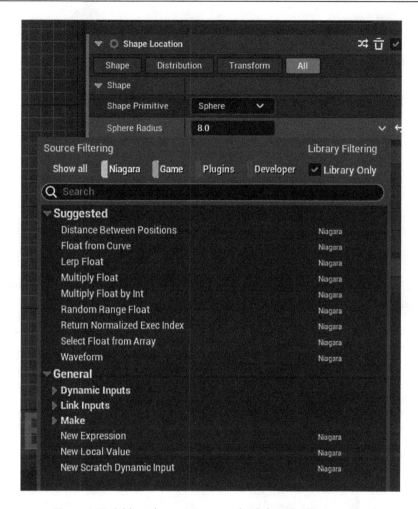

Figure 4.17: Adding dynamic input to the Sphere Radius property

We have now familiarized ourselves with the various features and functionality of the UI, including the different panels, and learned how to navigate the interface.

Let us get back to our emitter and understand it in more detail.

Creating an Emitter

Now that we understand the interface of the Niagara Editor, let us create an emitter step by step from existing templates and modify its properties so that it looks similar to *Figure 4.18*, the target effect that we want to achieve. This emitter consists of a stream of streaky particles that glow yellow when emitted and turn red as they age. The particles also emit light and are stretched in the direction of their motion. They will randomly change their movement path as if affected by turbulent wind.

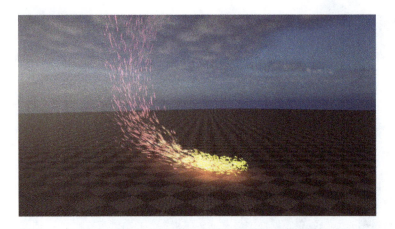

Figure 4.18: Let's create an emitter that looks like this

Let's start by creating a new emitter and call it `FireSparks`. Follow the instructions given in the previous section to create an emitter and then choose the fountain template.

You will end up with an emitter overview node similar to *Figure 4.19* with the white-colored particle fountain.

Figure 4.19: Our FireSparks emitter node

We are going to modify this emitter to achieve our target effect.

Taking a step toward achieving the target effect

Let us begin our journey toward creating the target effect.

We will start by changing the **Shape Location** module from **Sphere** to **Torus**:

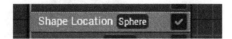

Figure 4.20 a: Select the Shape Location module

1. To do so, select the **Shape Location** module in the **Overview** node by single-clicking on it.
2. In the **Selection** panel, you will see the corresponding details show up. In the **Shape Primitive** property, choose **Torus** from the drop-down menu.

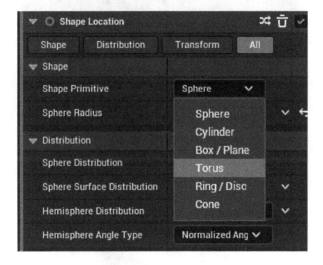

Figure 4.20 b: Select the Torus option

If you follow the preceding steps, the properties in the **Selection** panel will change to give you access to the **Torus** properties (*Figure 4.20 c*).

Figure 4.20 c: The Shape Location properties change to show Torus-related properties

Let's keep all the values at their default. Feel free to change the values as per your liking later.

Figure 4.21: The preview will change to show the fountain having a wider base

You will see that the fountain now has a wider base compared to that in *Figure 4.19*.

Enabling the fountain's movement in a random direction

We will now disable the **Gravity Force** module and add a **Curl Noise Force** module (*Figure 4.23*). This will make the particle movement a bit turbulent.

To add the **Curl Noise** module, click on the green + icon next to the **Particle Update** group panel. The **Add new Module** popup will appear. In that popup, search for **Curl Noise Force**, as shown in *Figure 4.22*.

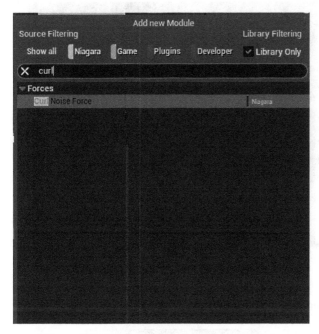

Figure 4.22: Adding the Curl Noise Force module

The **Curl Noise Force** module will be added to the **Overview** node, as shown in *Figure 4.22*.

Figure 4.23: Curl Noise Force module added to the Overview node

For **Curl Noise Force**, set **Noise Strength** to 10000.0 and **Noise Frequency** to 1.0. This will distort the flow of the particles and make them seem turbulent. However, the turbulence will seem locked in place and the particle movement will not feel truly random.

Now, check **Pan Noise Field** and set the value of **Y** to 0.5. This will result in **Pan Noise Field** animating the noise field to move in the **Y** direction at a speed of 0.5. This animates the turbulent flow of the particles.

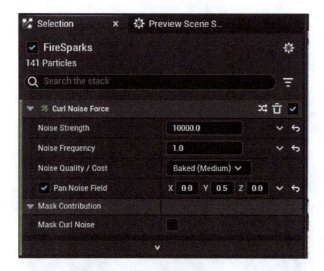

Figure 4.24: Changing the Curl Noise Force properties

You will see that the fountain is now moving in random directions.

Figure 4.25: Randomly moving particles affected by Curl Noise Force

You will notice, however, that the particles are still round shaped; we want the particles to stretch in the direction of the movement.

To do that, we have to make changes in two places.

Helping the particles stretch in the direction of movement

One of the two required changes is in the **Initialize Particle** module under the **Particle Spawn** section.

Here's how you do it.

Figure 4.26: The Sprite Size Mode property under Particle Spawn > Initialize Particle > Sprite Attributes

In the **Initialize Particle** module, go to the **Sprite Attributes** section and change **Sprite Size Mode** from **Random Uniform** to **Random Non-Uniform**.

This will change the entries below it from **Uniform Sprite Size Min** and **Uniform Sprite Size Max** to **Sprite Size Min** and **Sprite Size Max** with independent inputs for the **X** axis and the **Y** axis.

Set **Sprite Size Min** to **X**: 5.0 and **Y**: 10.0 and set **Sprite Size Max** to **X**: 5.0 and **Y**: 30.0.

Figure 4.27: Giving the particle sprites random non-uniform sizes

This will cause our sprites to be elongated but in a random direction and not in the direction of the particle velocity as we wanted.

Figure 4.28: The preview window showing our randomly elongated particles

Now for the second change. To set the particles to stretch in the direction of particle movement, select the **Sprite Renderer** module, which is under the **Render** group.

Figure 4.29: The Sprite Renderer module

Here, under the **Sprite Rendering** group, set the **Alignment** property from **Unaligned** to **Velocity Aligned**, as you can see in *Figure 4.30*.

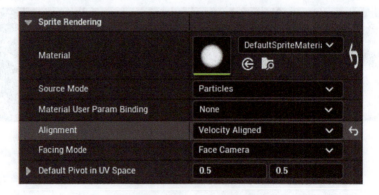

Figure 4.30: Changing Alignment to Velocity Aligned

You will see in the preview window that the particles are now properly stretched.

Figure 4.31: The particles are now properly stretched in the direction of movement

Since the number of particles is sparse, let's increase them by increasing **Spawn Rate**, which is under the **Emitter Update** group.

Figure 4.32: The Spawn Rate module under Emitter Update

Let's increase **Spawn Rate** from 90 to 2000.

Figure 4.33: The emitter preview shows a lot more particles due to a higher spawn rate

We will now have a lot of particles, as shown in *Figure 4.33*.

Changing the colors of the sprites

We now have to change the color of the sprites. We would like the sprites to be yellow at the start and then slowly turn red. This can be done by modifying the **Scale Color** module, which is under the **Particle Update** group.

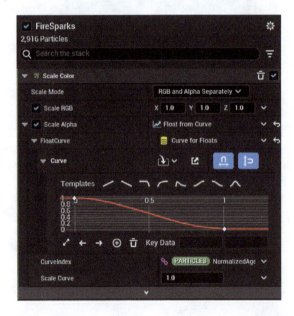

Figure 4.34: The Scale Color module the way it is set up in the fountain template

We will modify this module to suit our requirements.

The **Scale Color** module is using **Scale Alpha** connected to a curve through a **Float from Curve** dynamic input to modify the alpha of the particles over its Normalized Age. The Normalized Age is obtained by scaling the age of particles to be between 0 and 1. We will be learning about Dynamic Inputs in *Chapter 6*.

In the **Scale Color** module, first, let's change **Scale Mode** from **RGB and Alpha Separately** to **RGBA Linear Color Curve**.

Figure 4.35: Changing Scale Mode to RGBA Linear Color Curve

This will change the panel layout to show a color gradient band, as shown in *Figure 4.36*, which lets us change the color and alpha values. The top part of the gradient has color stops that we can edit by double-clicking on them.

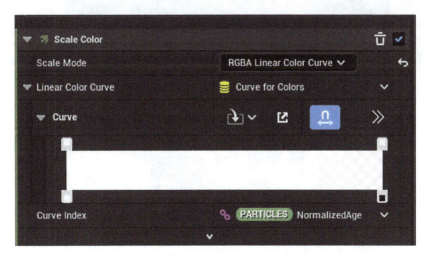

Figure 4.36: Linear Color Curve shown as a color/alpha gradient
with the color stops indicated by the arrows

Clicking on the color stop opens a color picker window where you can set the color.

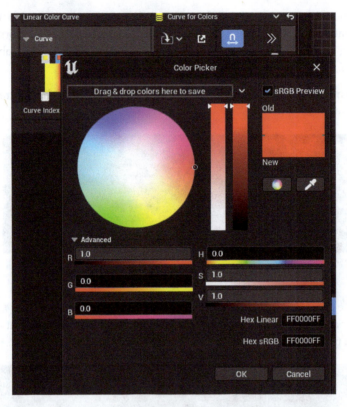

Figure 4.37: The color picker window that pops up when you click on the color stop

The bottom-part color/alpha gradient has the alpha stops. Click on the alpha stops to set the alpha values of the gradient.

Figure 4.38: The alpha stops to set the opacity

Modify the gradient so that it looks like *Figure 4.39*.

Figure 4.39: Setting the gradient on Linear Color Curve

In the color picker, increase the value of the **V** slider to 500.0. This will give us an emissive shader that will make the particles glow.

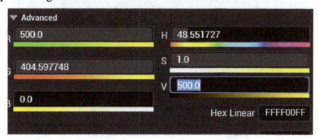

Figure 4.40: Setting the V amount to 500.0

Our particle emitter is now very close to our target look.

Figure 4.41: The particles are now starting to look like fiery sparks

We have a few things remaining. First, we have to get the particles to collide with the floor, and second, we need to get the particles to emit light.

Enabling collision with the floor

Let's start by enabling collision with the floor. Before we enable the collision, we want the floor to be enabled in the preview window. To do that, open the **Preview Scene Settings** tab by checking the option in the **Window** menu.

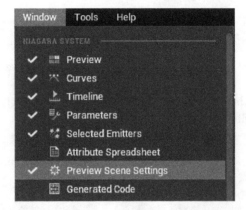

Figure 4.42: Accessing the Preview Scene Settings tab

In the **Preview Scene Settings** tab that has now opened, check **Show Floor**.

You should now see the floor in the **Preview** panel. Once we enable collision, our particles can bounce off this floor.

Figure 4.43: Enabling Show Floor in the Preview Scene Settings tab

To enable collision, click on the green plus button in the **Particle Update** group label and add the **Collision** module to the emitter node.

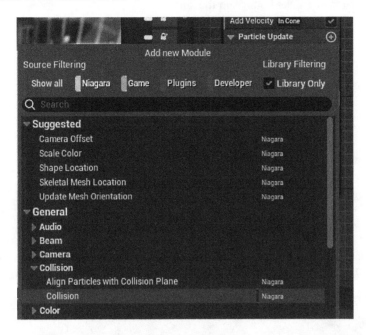

Figure 4.44: Adding the Collision module to the Particle Update section

To make it easy to find any modules in the **Add new Module** popup, type the module name in the **Search** box. Ticking the **Library Only** checkbox ensures that only modules added to the library are shown.

Figure 4.45: Collision module added

You don't need to change any settings in the **Collision** module, as the default settings will be good enough.

Now you will see that the particles bounce on the floor.

Figure 4.46 a: Particles bouncing off the floor

If your particles don't seem to be bouncing off the floor in the **Preview** panel, add a little Z offset to the particle in the **Initialize Particle** module by checking on **Position Offset** and setting **Z** to 50.0.

Figure 4.46 b: Position offset set to 50.0

This emitter is almost there. We will now make the particles cast light. This will make the particles look natural as glowing hot sparks cast light on surfaces around them.

Next, we'll see how to do that.

Making the particles cast light

For this, we have to add a new type of Renderer to the **Emitter** node. We can add a new Renderer by clicking on the red plus sign next to the **Render** group label. The Renderer we are going to add is **Light Renderer**.

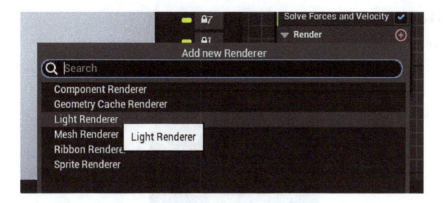

Figure 4.47: Adding the Light Renderer module

The **Light Renderer** module is added in addition to the existing **Sprite Renderer** module.

Figure 4.48: The Light Renderer module along with the Sprite Renderer module

The particles will now start casting a small pool of light around them.

If you want to increase the light intensity of the particles, increase the values of the vector in **Color Add** and **Radius Scale** in the **Light Renderer** properties. This will make the lighting appear more distinct.

Figure 4.49: The resulting render in the Preview panel with particles casting light around them

We are now done with our emitter.

The **Emitter** node should look like this:

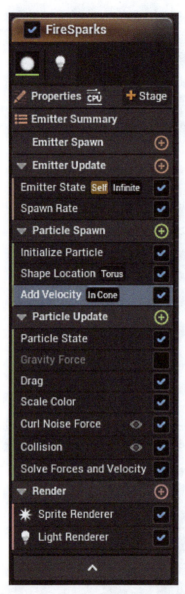

Figure 4.50: The completed emitter node

We are now in for a rude surprise. If we try to drag the emitter into our scene, we will see that Unreal Engine does not allow us to do so.

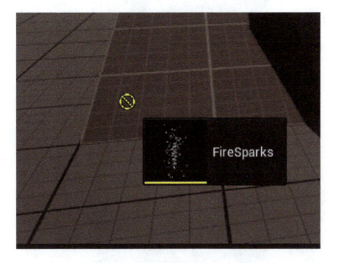

Figure 4.51: Emitter node cannot be added to the level

This is because you can only add a Niagara System to a level. We will need to create a Niagara System that contains the emitter we just created.

Creating a Niagara System

Let's create a Niagara System.

This can be done by right-clicking in an empty place in the Content Browser and choosing **Niagara System** from the menu.

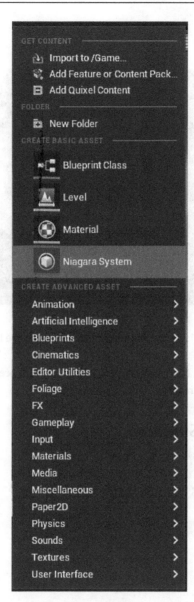

Figure 4.52: Creating a Niagara System directly through the right-click menu

Optionally, you could also go to the **FX** submenu and choose **Niagara System**.

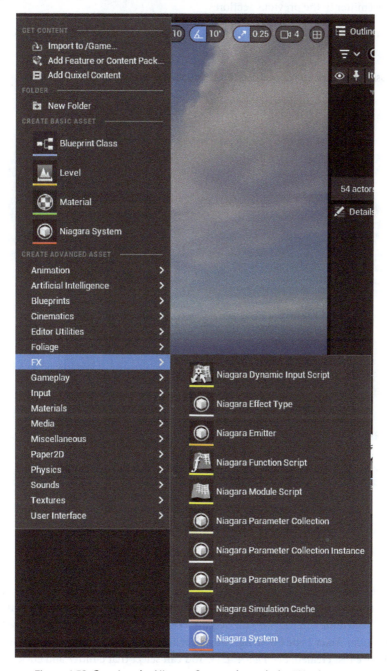

Figure 4.53: Creating the Niagara System through the FX submenu

Choosing the **Niagara System** option pops up a new wizard window similar to the one we had when we created a new emitter in the previous section.

Choose the first option, **New system from selected emitter(s)**. Click the **Next** button to move to the next screen.

We will be choosing the **Fire Sparks** emitter on the next screen.

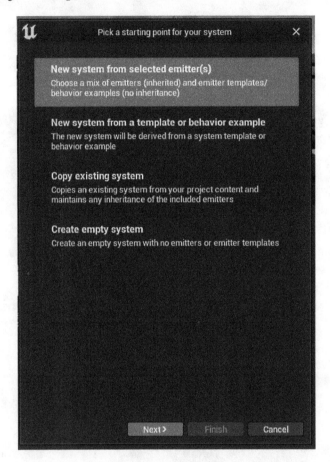

Figure 4.54: The Niagara System wizard

On the next screen, we will see a layout similar to *Figure 4.5* that we saw for our emitter. This can be used to directly create a system using one of the templates we use to create an emitter. However, since we have already created our emitter with the required customization, we need to choose the **Parent Emitters** option under **Asset Type Filtering**. After selecting the **Parent Emitters** option, you will find listed the emitter we just created. See *Figure 4.55*. You can select that emitter and click on the green + button at the bottom of the pop-up window.

If you wish, you can add additional emitters from the **Templates** filter option.

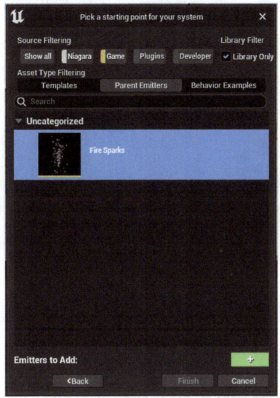

Figure 4.55: Our Fire Sparks emitter is listed under Parent Emitters

Our Niagara System is now created, and you will see the Niagara System asset icon in the Content Browser.

Figure 4.56: The Niagara System asset icon

To edit the Niagara System, double-click on the icon. This will open the Niagara System Editor. The node panel has the word **SYSTEM** written in a large font to help differentiate the window from the emitter window.

Figure 4.57: The Niagara System Editor

The Niagara System Editor is very similar to the Niagara Emitter Editor. The difference is that you can now have multiple **Emitter** nodes and an extra blue accented **System** node. We will be exploring the System Hierarchy and Overrides in *Chapter 5*.

There is also a shortcut for creating a Niagara System from the emitter. You can right-click on the emitter and, from the menu that appears, choose **Create Niagara System**. You can bypass all the preceding steps with this method.

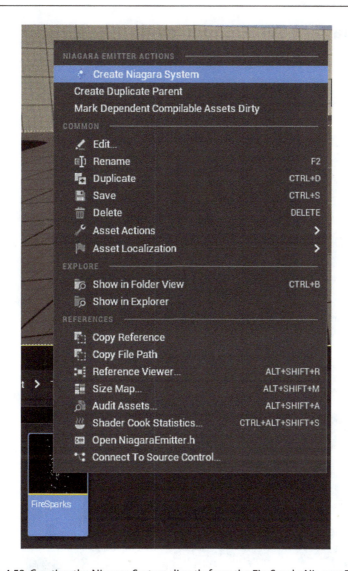

Figure 4.58: Creating the Niagara System directly from the FireSparks Niagara Emitter

Now that the Niagara System is ready, you can add it to the level.

Adding a Niagara System to a level

This is pretty straightforward. Just drag the Niagara System from the Content Browser into the level.

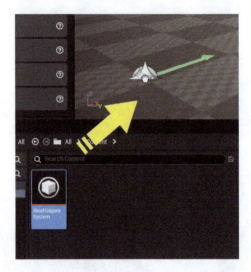

Figure 4.59: Dragging the Niagara System into the level

The Niagara System will start playing.

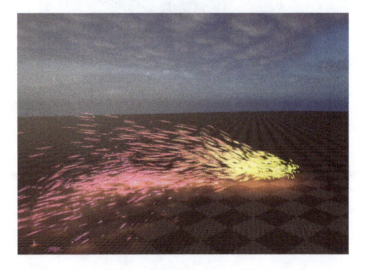

Figure 4.60: The final look of the Niagara System

The particle system should look like *Figure 4.60*.

Adding a Niagara System to a Blueprint Actor

We added a Niagara System to the level by dragging it into the level. But that is not the only way of adding a Niagara System to a level. Niagara Systems can also be added as part of a blueprint class actor. When you develop complex blueprint assets that have particle effects as a part of their creative design, it is preferable to just integrate a Niagara System into the Blueprint itself rather than keeping it as an individual asset to be dropped into a level. This will also allow us to build relationships between the behavior of a Blueprint and the behavior of a particle system. For example, we could have the particle system change color as the blueprint asset speeds up. For now, let us see how we can add a Niagara System to a blueprint class.

Let's create a Blueprint class that incorporates a Niagara System.

Right-click in a blank area in the Content Browser to open the menu and choose the **Blueprint Class** option.

Figure 4.61: Creating a blueprint class

In the subsequent window that opens, click on the **Actor** button to create an Actor Blueprint. Of course, you can use any of the other blueprint classes, such as **Pawn** and **Character**.

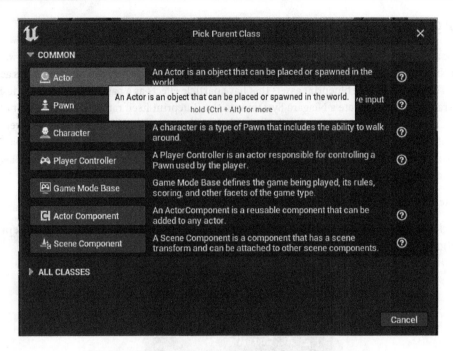

Figure 4.62: Choosing the parent class for our Blueprint

In the class we created, we will add **Niagara Particle System Component**. This can be done by clicking on the **Add** button in the **Components** panel and choosing **Niagara Particle System Component** from the menu that pops up. If you cannot find it right away, try using the **Search Components** box.

Figure 4.63: Adding Niagara particle system Component to our Blueprint

This adds a component named **Niagara**. Feel free to rename it to any suitable name by pressing *F2* or right-clicking it and choosing **Rename** in the menu.

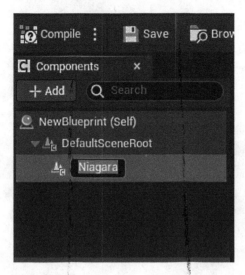

Figure 4.64: Niagara particle system Component added. Feel free to rename this component

Now, choose this component by single-clicking on it. This will show the details of this component in the **Details** panel. Under the **Niagara** section, you will have a property called **Niagara System Asset**. You can drag the Niagara System we created previously into this property slot. Our Blueprint Actor is now ready. You can drag the actor into the level to see it work.

Figure 4.65: Adding the Niagara System we created to the Niagara
System Asset property in the Details panel

In one of the upcoming chapters, we will see how we can use Blueprint public variables to change the properties in a Niagara particle system. This will allow us to define the Niagara particle system behavior based on other components and the blueprint's behavior. It also helps to make it easy for people not familiar with Niagara to modify the particle system's behavior indirectly through the public variables presented as abstracted (simplified) blueprint properties.

Summary

In this chapter, we created our first Niagara Particle Emitter asset as well as our first Niagara particle system asset. We also saw how the Niagara particle system asset references Emitter assets.

Finally, we learned how to add a Niagara component to a level and a blueprint class.

We will be exploring the relationship between the emitter and the system in *Chapter 5*, which is the next chapter.

5

Diving into Emitter-System Overrides

In this chapter, we will explore overrides in Niagara.

Overrides are used to extend or modify the behavior of existing objects in the system without making destructive changes to the original settings of the system, as well as allowing for customizations specific to a particular implementation. We will understand how to use overrides to develop a workflow where we can reuse our emitters in multiple projects.

The override workflow is useful as it extends the functionality of an existing particle system. This modified particle system can access the inputs and outputs of the original particle system and modify its behavior as needed. It also lets us design a production workflow where we can reuse existing particle systems and tweak them as required in our Unreal projects. We can have a library of basic effects and then tweak them on a per-project basis.

After that, we will also learn about emitter and module defaults and how they help while developing particle effects.

In this chapter, we will be covering the following:

- Module Override
- Parameter Override

Technical requirements

Unreal Engine 5.1 or above is required. The installation procedure was explained in *Chapter 1*.

You can find the project we worked on in this book here:

```
https://github.com/PacktPublishing/Build-Stunning-Real-time-VFX-
with-Unreal-Engine-5
```

Module Override

Niagara offers a unique workflow whereby one can create a library of emitters with generic values and modules and then tweak them as per specific project requirements. This is possible because of the override functionality. Module overrides allow users to extend the behavior of existing modules by adding override modules to the emitters. These override modules are added at the Particle System level. Let us understand this with a practical example. In the previous chapters, we created an emitter, and we then called the emitter in the Niagara System.

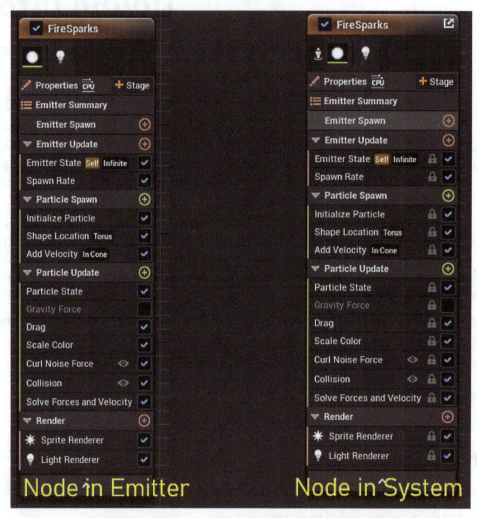

Figure 5.1: Comparing the Overview nodes for the same emitter

Let us compare how the **Overview** nodes differ when they are an Emitter asset and when they are a Particle System asset. We know that the Emitter asset is referenced in the Particle System asset, and we expect the same to be the case for the **Overview** nodes. Let us see whether there are any differences. When you compare the **Overview** node of **FireSparks** in the emitter with the **Overview** node of **FireSparks** in the System, you will find that apart from the **Enable Isolation** icon and the lock icon on the **System** node, both nodes look almost exactly the same. Once we have added our override modules, which we will do later in this chapter, we will see that these nodes will be different from each other.

The **System** node is driven by the **Emitter** node and any changes made and saved to the **Emitter** node will be reflected in the **System** node.

To demonstrate that, let's try to remove the disabled **Gravity Force** module from the **Emitter** node. You can select the **Gravity Force** module and either press *Delete* on your keyboard or click on the trash can at the top right of the **Gravity** section in the selection panel.

Figure 5.2: Deleting the Gravity Force module in the Emitter
Overview node by clicking on the trash can icon

Before you click on the trash can icon in the emitter, let us check whether a similar icon exists on the **Gravity** module in the **System** node. You will find that in the **System** node, if you select the **Gravity** module, instead of the trash can icon, you will find a lock icon.

Figure 5.3: The lock icon on the Gravity Force module in the System node

This is an important difference; let us explore it a bit more.

Another way to remove a module in a node is by selecting the module in the **Overview** node and pressing the *Delete* key on your keyboard.

In the **Emitter Overview** node, select the **Gravity** module and delete it. It will disappear without any problems. The **Gravity** module will still be there in the **System** node until you save the emitter. Once you save the emitter, the **System** nodes will update, and the **Gravity** module will disappear from the **System** node too.

This was expected as we know that the **System** nodes read from the emitter nodes.

Let's undo the deletion of the **Gravity** module by pressing *Ctrl + Z* so that the **Gravity** module is back in both the System and the emitter.

Now, let's try to delete the **Gravity** module again, but this time we will try to delete it not in the emitter but in the system. You will get a notification that it cannot be deleted.

> **Fun fact!**
> This notification contains a typo, which has gone unnoticed for many versions of Niagara.

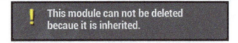

Figure 5.4: The infamous "becaue" typo

This makes it clear that while you can add/remove modules in the emitter and the changes you make will reflect in the System, you cannot remove inherited modules added in the emitter from the System.

You can, however, enable or disable the module. In the following screenshot, we see that we can enable the **Gravity** module in the System, although we have disabled it in the Emitter.

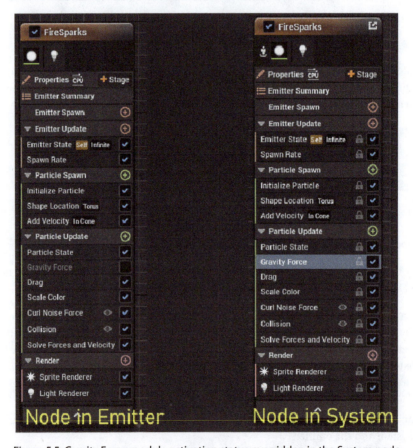

Figure 5.5: Gravity Force module activation status overridden in the System node

You can add additional modules in the **System** node over and above the modules inherited by it. In our case, let's add a **Point Attraction Force** to our **System** node in the **Particle Update** section. You can do this by clicking on the green + icon on the right of the **Particle Update** section panel, then searching for the **Point Attraction Force** item in the resulting pop-up menu, and finally clicking it.

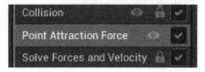

Figure 5.6: The Point Attraction Force module added to the Particle Update section

You might not see any difference in behavior in the preview window and that is because the default values in the module are very low. Let's modify some of the properties. Let us change **Attraction Strength** to 1000 and **Attraction Radius** to 500.

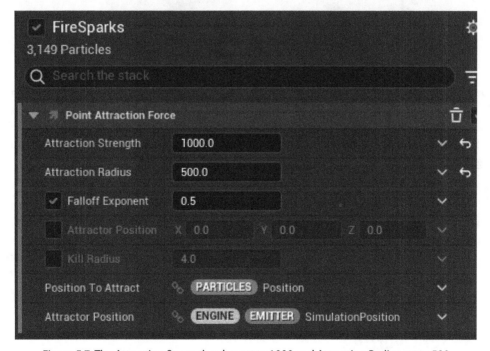

Figure 5.7: The Attraction Strength value set to 1000 and Attraction Radius set to 500

The particle system behavior in the level will change in response to this.

Figure 5.8: The modified behavior where the particles are attracted to the center of the particle system

What is to be noted is that if you save the System, you will see that the added node is present only in the System and not in the Emitter. We are thus overriding the module structure of the Emitter by adding additional modules in the **System** nodes.

The advantage of adding modules in the System rather than in the Emitter is that we can delete them from the System if required, unlike the previous case where if a module was inherited from the Emitter, we were not able to delete it. This allows us to have a library of emitters with the basic modules that we can then tweak in the System according to specific project requirements. Now that we understand module override in a System, let us see what happens when we delete a module. Let us delete **Point Attraction Force**. You can do this by either clicking on the trash can icon in the details panel or selecting the module and pressing the *Delete* key. As expected, the module will be deleted without any issues. Let's save the System before moving ahead.

Parameter Override

Sometimes we need to override only particular properties of a module rather than the whole module at the System level. Niagara allows us to do that too. This helps us create a library of modules for reuse and not have to worry about the correct values for all parameters. These parameters will be tweaked in the **System** node by overriding the values set in the emitter. Let us now see how the override functionality works when it comes to module parameter values.

First, we will select the **Emitter** node and select the **Curl Noise Force** module. We see that for the **Curl Noise Force** module, we have set the value of **Noise Frequency** to 1.0. This is not the default value of this property. Therefore, a white arrow pointing left will be shown on the right side of the property.

Figure 5.9: White reset arrow in the Emitter node property

Clicking on that arrow will reset the value of the property to its module default of **50.0**. Once the value is set to the module default, the arrow disappears. A module default value is set internally in the module as a default value.

Figure 5.10: The value of Noise Frequency defaults to 50.0 when the white arrow is pressed

If we change the property again, the arrow reappears. Let's set the value of **Noise Frequency** to 1.0 as it was before and have the arrow return.

Now that we have seen the behavior in the **Emitter** node, let's see how the override behavior works in the **System** node.

In the **System** node, as you would expect, we have **Noise Frequency** set to 1.0 because that is the value that we set in the **Emitter** node from where it was inherited. There is a corresponding white arrow that will reset the value to the module default.

Figure 5.11: The white arrow shows up in the System node's Noise
Frequency property when the value is set to 1.0

But there is a little difference in behavior here. Clicking on the arrow will reset the default value to **50.0** but watch what happens to the arrow. Instead of the arrow just disappearing, now you have a green arrow appearing just next to where the white arrow used to be. This arrow resets the value of the property to its Emitter default. The Emitter default is the value set in the emitter from which we have inherited this node, which was 1.0. But as we know, setting it to 1.0 means it is now not equal to the module default value of **50.0**. So, if you click on the green arrow, the value will be reset to the emitter default value of 1.0 but the white arrow will now be back as the value is not equal to the module default.

Figure 5.12: A green reset arrow appears on clicking the white
reset arrow and the property value is set to 50.0

These arrows allow us to choose between the module default or the emitter default as a value to reset to should we want to start again while we are experimenting with the values. The green arrow resets the value to the emitter default while the white arrow resets the value to the module default.

> **Historical fact**
> The module default arrow used to be yellow in Unreal Engine 4!

The System value will be the one that overrides any inherited values from the emitter or the module. We can easily access the default values set up in the emitter or the module with the help of the arrows.

This feature enables us to design a workflow where the emitter has its module values set to an amount that supports a broad variety of effects, and then at the system level, we tweak the values to adapt them to our specific target effect.

The **Override** feature provides a flexible and maintainable way to customize the behavior of particle systems in the Niagara framework.

These override capabilities enable us to implement a team workflow where part of the team that is good at Niagara develops the Niagara emitter modules that are not designed for specific use cases but serve as a library of effects to be used by other team members. The rest of the team members can add the original emitter modules to their Niagara Systems and tweak them in the system modules to achieve the effect that they want without modifying the original effect modules.

Summary

In this chapter, we saw the relationship between **Emitter** nodes and **System** nodes. We saw how we can add and override modules in the **System** node. We also saw how specific parameters in a module can be overridden in the **System** node. We understood that this enables us to design a workflow where a team can develop a library of emitter modules for reuse in all their projects and tweak the values of the emitters in the **System** node.

In the next chapter, we will learn about Dynamic Inputs.

Part 2: Dive Deeper into Niagara for VFX

With our fundamentals now in place, in this part, we will start exploring advanced concepts of Niagara. We will create custom behaviors by creating our own custom modules and Scratch modules. We will learn about the interaction between emitters using Events and Event Handlers. We will learn about debugging and optimizing our particle systems, and finally, we will create a Blueprint Actor containing a Niagara system controlled by public variables.

This section comprises the following chapters:

- *Chapter 6, Exploring Dynamic Inputs*
- *Chapter 7, Creating Custom Niagara Modules*
- *Chapter 8, Local Modules and Versioning*
- *Chapter 9, Events and Event Handlers*
- *Chapter 10, Debugging Workflow in Niagara*
- *Chapter 11, Controlling Niagara Particles Using Blueprints*

6

Exploring Dynamic Inputs

Now that we have explored modules, as the next step, we will be exploring the concept of Dynamic Inputs in this chapter. With Dynamic Inputs, you will be able to create more interesting particle effects than would have been possible by using the standard inputs available in modules.

In this chapter, we'll be covering the following topics:

- What are Dynamic Inputs?
- Creating random colored particles using Dynamic Inputs

Technical requirements

You can find the project we worked on in this book here:

```
https://github.com/PacktPublishing/Build-Stunning-Real-time-VFX-
with-Unreal-Engine-5
```

What are Dynamic Inputs?

Niagara is designed to enable anybody to work with it regardless of their level of knowledge about it. Everything that we learned about Niagara in the previous chapters enables us to create relatively complex particle systems. However, it is pretty obvious that sooner or later, an artist will want more extensibility. This can be achieved by having property values that are not constant but are a result of some mathematical formula or function. Niagara provides a library of such formulas and functions, which we call **Dynamic Inputs**. Dynamic Inputs enable users to have infinite extensibility.

Dynamic Input can drive any value using different types of functions or graph logic. Dynamic Inputs can also be chained to create more complex particle effects than would be possible with just one Dynamic Input. It is advised to use Dynamic Inputs with existing modules whenever possible, rather than creating new modules to improve performance.

Dynamic Inputs can be added to almost any module parameter. The parameter to which Dynamic Inputs can be added will have a downward-pointing arrow on the right-hand side, next to the **Reset to Default** arrow (as you can see in the following figure).

Figure 6.1: The Dynamic Input icon on the right side of a module property

Clicking on the arrow will reveal a context-sensitive menu of viable Dynamic Inputs.

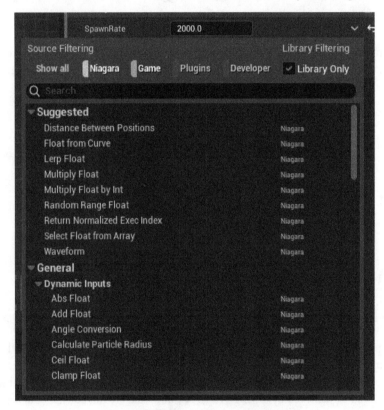

Figure 6.2: The Dynamic Input popup

By default, in the pop-up menu, Niagara hides any custom Dynamic Inputs added by you and any plugins that you may have. You can enable them by clicking on the **Plugins** and **Developer** selection option or by clicking on the **Show all** button.

Niagara also tries to make things easier for learners by suggesting a few Dynamic Inputs. The suggested inputs are inputs that you are most likely looking for. They help save the time that is usually spent searching for them.

Now, after that brief introduction, let us see an example of how to use Dynamic Inputs. Open the **NewNiagaraSystem** particle system containing the **FireSparks** emitter that we created in *Chapter 4*. We will modify a few properties by adding Dynamic Inputs to them. In the **Overview** node, let us choose the **SpawnRate** property under **Emitter Update**. It should have a spawn rate of 2 , 000. Let's use Dynamic Inputs to make this vary across time, just as an example.

To do that, select the **SpawnRate** module, and in the **Selection** panel, click on the downward-pointing arrow. From the **Suggested** options, choose **Waveform**.

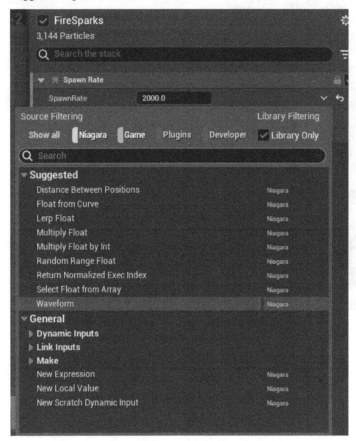

Figure 6.3: Adding the Waveform Dynamic Input

This will add the **Waveform** Dynamic Input properties to the **SpawnRate** property. This will be indicated by a blue graph icon next to the word **Waveform**, which would have replaced the text input widget, which was **2000.0** in *Figure 6.1*. A **Waveform** input applies different waveforms, such as sine, cosine, square, and triangle. For each of these waveforms, we can modify the parameters to get different behaviors. We can also apply multiple waveforms to get the resulting interference behavior. In our case, let us keep a single waveform, which will be the default **Sine** waveform.

SpawnRate is now being driven by the **Waveform** Dynamic Input instead of being set to 2000.0. You may be a bit confused as you may not see any significant changes in the particle behavior.

Figure 6.4: The Waveform Dynamic Input properties

The effect of the Dynamic Input will be evident after we tweak some of its values to significantly influence our particle system. Let's make the following changes:

- Set **Global Amplitude Scale** to 500.0
- Set **Amplitude Min/Max X** to -1.0

This will scale up the **Sine** waveform to a sufficient value so that we will be able to see **SpawnRate** increase and decrease in our particle system. As **Amplitude** approaches -1, the particles stop spawning and start increasing. Try playing around with **Global Amplitude Scale** and take it up to 5000.0 to see how the particles react. With a value of 5000.0, you will see a whole lot of particles being spawned as the sine value approaches 1.

We have just created a pulsating spark generator using Dynamic Inputs.

Play around with the other properties to see how they affect the particle system.

Let us now reset **SpawnRate** to a single value as we had before we added the **Waveform** Dynamic Input. To do that, click on the downward-pointing arrow to reveal the Dynamic Input. In the **Make** section, choose **New Local Value**, and **SpawnRate** will now be driven by the same number as before.

Figure 6.5: Removing the Dynamic Input

We saw a simple Dynamic Input create a more interesting effect than would otherwise be possible by only using modules.

Next, we will see how to chain Dynamic Inputs to create even more interesting effects.

Creating random colored particles using Dynamic Inputs

In this section, we will see how we can chain Dynamic Inputs to create randomly colored particles.

We'll continue to use the **NewNiagaraSystem** particle system. We will also explore a few new UI elements added to the **Selection** panel that make the process a bit easier.

Let us start by double-clicking on **NewNiagaraSystem** to open it in the Niagara Editor. First, we need to disable a few modules in our System to prevent conflicts and enable our Dynamic Inputs to work properly. The module that we will disable is the **Scale Color** module in the **Particle Update** section. This module would otherwise overwhelm the random colors assigned by our Dynamic Inputs. There is no technical requirement for us to disable the module. We have only done so to make it easier to understand our Dynamic Input.

Figure 6.6: Disabling the Scale Color module

After disabling the **Scale Color** module, all the particles will be colored white and lose their glow. We have also left the previously disabled **Gravity Force** as it is to let the particles float. We are now ready to create our randomly colored particles.

The result we expect is that our particles will be assigned a random color at spawn time and will continue to have that color until they die. This means that the particles would need to be assigned a color in the **Particle Spawn** group. We already have the appropriate module, that is, the **Initialize Particle** module. The **Initialize Particle** module applies properties specified in the module to particles at spawn time.

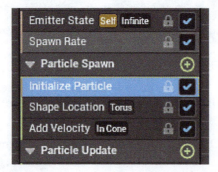

Figure 6.7: The Initialize Particle module

Since we want to modify the color of the particles, we will focus on the **Color** property in the module.

Figure 6.8: The Color property in the Initialize Particle module

You will notice a **Color Mode** property, which we will come back to later. For now, let us just focus on the **Color** property.

By default, the **Color** property has a **Linear Color** value assigned to it. We want a Dynamic Input that assigns a random linear color to this property.

Figure 6.9: Color has a value of type Linear Color

Let us check what Dynamic Inputs we have available. Click on the downward-pointing arrow to show the Dynamic Inputs menu.

Niagara will helpfully suggest a few options. Since we need a random linear color, choose **Random Range Linear Color** from the suggested options.

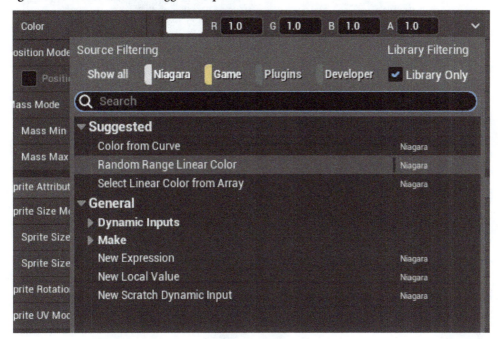

Figure 6.10: Choose the Random Range Linear Color property

The **Color** property will now show the Dynamic Input, as shown in *Figure 6.10*, where you can choose two colors that will define the range from which a random color is chosen. It is a bit difficult to see the variations in gray from the default minimum black color and the maximum white color, so let us set the **Minimum** color to red and the **Maximum** color to green.

Figure 6.11: Setting the Minimum and Maximum color ranges in Random Range Linear Color

This will result in particles getting a random color between red and green assigned to them, as seen in *Figure 6.12*.

Figure 6.12: Particles with a random color between red and green as seen in the Preview window

This is not enough to get multicolored particles. So, we need to chain this Dynamic Input.

Chaining Dynamic Input means adding a Dynamic Input to an existing Dynamic Input.

In this case, we will take the **Maximum** value in the **Random Range Linear Color** Dynamic Input and add another **Random Range Linear Color** input to it. We can do this by clicking on the downward-pointing arrow on the right of the **Maximum** property, which we had set to green.

Figure 6.13: Chaining another Random Range Linear Color property to the
Maximum property of the first Random Range Linear Color property

You will now have the **Maximum** value replaced by a **Random Range Linear Color** input, which will give you an additional two colors to choose from. Let us set **Minimum**, in this case, to green and **Maximum** to blue.

Figure 6.14: The Color property with two Random Range Linear Color Dynamic Inputs chained together

The Dynamic Inputs will now result in our particle system being able to choose from the whole spectrum of RGB colors. This will result in truly random colors being assigned to the particles.

Figure 6.15: Random color with all colors from the RGB spectrum as a result
of chaining two Random Range Linear Color Dynamic Inputs

This is a simple example of chained Dynamic Inputs; you can use chaining to create complex Dynamic Inputs for your particle system.

This method, while it works, is a bit too cumbersome to achieve something as simple as assigning random colors to a particle system.

For such common behaviors, Niagara has another, easier method. This is through the **Color Mode** option that we skipped earlier. **Color Mode** gives the user an easier workflow to achieve variations in **Random Hue/Saturation/Value** and **Alpha** for the particle system. Let us achieve the same behavior of random multicolored particles using the **Color Mode** option. Before we do that, let's remove the **Random Range Linear Color** Dynamic Input by choosing **New Local Value** from the Dynamic Input menu. We should be back to having just one choice of color.

Color Mode would be set to **Direct Set**. Let us choose the **Random Range** option from the **Color Mode** drop-down menu.

Figure 6.16: Choosing Random Range's Color Mode

This will change the **Color** property to allow you to choose two colors, just like we did with the **Random Range Linear Color** Dynamic Input. Let us set it to red and green to achieve a similar behavior as earlier.

Figure 6.17: The UI changes to show Color Minimum and Color Maximum
as properties when Random Range is chosen in Color Mode

Similar to the situation we faced earlier, we just get colors between red and green and not truly multicolored particles. Now, we could add **Random Range Linear Color** to either the **Color Minimum** or **Color Maximum** property to get the randomly colored particles, but Niagara has made the process a bit easier than that.

To get truly randomly colored particles, choose the **Random Hue/Saturation/Value** option in the **Color Mode** property.

Figure 6.18: Choosing Random Hue/Saturation/Value under Color Mode

This will change the **Color** property UI and give you additional options. You will be able to define a hue variation range for your particle color randomness. Make sure to choose a color other than pure white or black to be able to see the variations.

In this case, we have chosen a shade of red, and we have set **Hue Shift Range** to **X** = -0.1 and **Y** = 0.1.

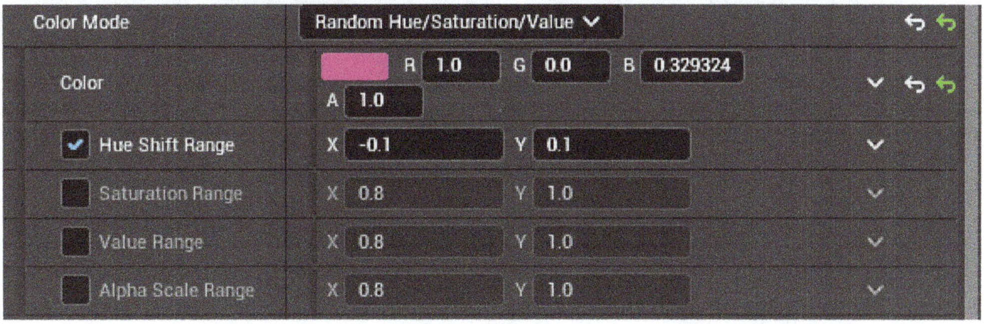

Figure 6.19: Choosing a shade of red with Hue Shift Range values of X = -0.1 and Y = 0.1

Making the previously mentioned changes will give us randomly colored particles that are red and shades near red, as seen in *Figure 6.20*.

Figure 6.20: Red-colored particles with randomly shifted hues within the specified range

To get truly multicolored particles encompassing the whole color spectrum, set **Hue Shift Range**'s **X** value to -0.5 and **Y** value to 0.5. This will extend the random hue range across the hue spectrum and give truly random colors.

Figure 6.21: Randomly colored particles across the full hue range

You can now tweak the colors of the particles by changing the values for **Saturation Range**, **Value Range**, and **Alpha Scale Range**. Unlike the hue range, you have to adjust the range from 0 to 1 as saturation, value, and alpha are normalized. You can also overdrive value and saturation to get emissive and highly saturated particles.

Summary

In this chapter, we learned how to use Dynamic Inputs. This enabled us to achieve even more diverse behaviors in our particle systems than would have been possible with just constant value inputs. We encourage you to experiment with the rest of the available Dynamic Inputs to see what you can achieve.

In the next chapter, we'll start looking at more advanced features. We have only used Niagara-provided modules before now. We will start to understand how to create custom Niagara modules.

7
Creating Custom Niagara Modules

Modules in Niagara enable us to achieve different behaviors for our particle systems. We are now familiar with a few of these modules. However, as you start working on advanced particle systems, it may not be possible to achieve the behavior that you require in your particle systems with the modules that are shipped with Niagara. This is when you may want to design your own custom modules. In this chapter, you will learn how to create your own custom Niagara modules.

The topics we will cover in this chapter are the following:

- Creating a new module
- The Niagara Module Script Editor
- Editing a Niagara module to create custom effects

Technical requirements

You can find the project we worked on in this book here:

https://github.com/PacktPublishing/Build-Stunning-Real-time-VFX-with-Unreal-Engine-5

Creating a new module

So far, we have been using inbuilt modules to create our particle effects. The real power of Niagara is unleashed when you start developing custom modules. Custom modules give you unlimited control over all aspects of your particle system. They allow you to design custom behavior for your particle system. Developing these modules requires some basic knowledge of mathematics and vectors. We covered some of the basics at the beginning of the book.

We will be creating a particle system with a custom module that detects the presence of the player and changes the size and color of particles around the player. As the player walks around the particle system, the change in color and size will follow the player. It is not possible to create this behavior by using just the inbuilt modules. To achieve this behavior, we will need to write a script in our custom module.

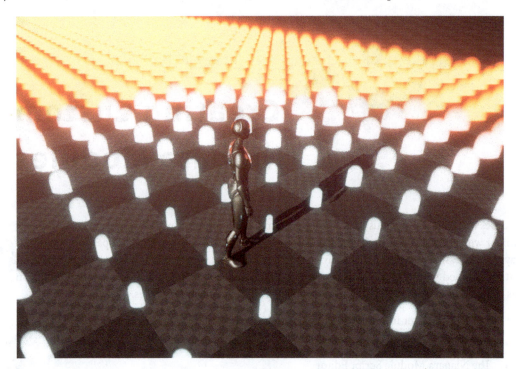

Figure 7.1: The effect we will have achieved by the end of this chapter

While it might be possible to create this effect using inbuilt modules, it would be more manageable to write a custom module for this. Not only would creating a custom module result in a more compact **Overview** node but it would also allow us to design an easier interface for the artist to tweak parameters. An alternative approach would be to create a Local module (Scratch module), which we will see later in the book. Developing a custom module is preferable if you expect to reuse the module in upcoming projects. A Scratch module is embedded into the particle system and is inconvenient to reuse in other projects' particle systems. A custom module, on the other hand, is saved as a separate asset and therefore can be used in other particle systems or shared with other users and projects.

A team can develop custom modules over the development cycle of its first game and can then reuse them in all its subsequent projects. The custom modules also can work with the Dynamic Inputs we learned about earlier, letting the user tweak the modules just like they would any inbuilt module.

Modules have their own editor with a workflow similar to blueprints with Niagara-specific nodes. The nodes can become very complex and can even have **High-Level Shader Language** (HLSL) scripts embedded into them for further customization. HLSL is the C-like language that you use with programmable shaders and is beyond the scope of this book.

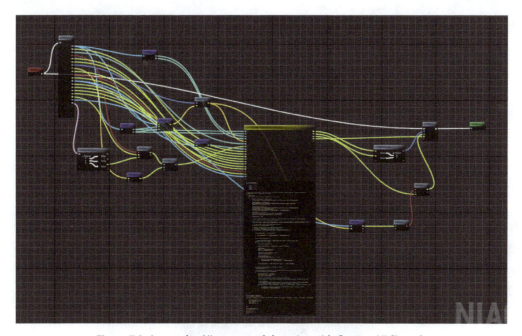

Figure 7.2: A complex Niagara module script with Custom HLSL node

Now that you know all about the importance of custom modules, let us start creating our very own custom module:

1. To create a custom Niagara module, right-click on **Content Browser** and choose **FX** > **Niagara Module Script**.

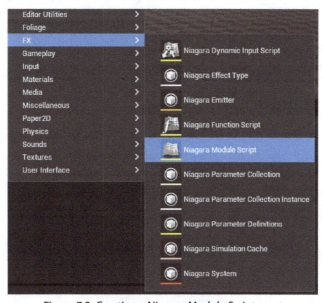

Figure 7.3: Creating a Niagara Module Script asset

2. This will create a **Niagara Module Script** asset. A **Niagara Module Script** asset has a yellow line under it. Let us rename this module `Presence Detector`.

Figure 7.4: The Niagara Module Script asset in Content Browser

This name is used as a label when we add this asset to the Niagara **Emitter/System** node.

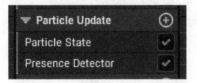

Figure 7.5: The Presence Detector Niagara Module Script asset
as a module entry in the Niagara System node

Next, let us begin editing the module. As you would expect, you must double-click on the **Niagara Module Script** asset to launch the Niagara Script Editor.

The Niagara Module Script Editor

The Niagara Module Script Editor looks suspiciously similar to a blueprint editor. This makes it easy for users familiar with blueprints to grasp the functionality easily. The nodes available here are completely different from what you otherwise see in a blueprint, although there may be a few nodes, especially the ones dealing with basic mathematics, that look similar.

Figure 7.6: The Niagara Module Script Editor

There are eight major areas in the editor.

Figure 7.7: The Menu bar

- The **Menu** bar

  This area contains commands to open/save the module, undo/redo options, a few asset commands, and commands to hide/show the different panels in the editor.

Figure 7.8: The Toolbar area

- **Toolbar**

  Along with the standard buttons to save and browse for the asset, the **Toolbar** area has buttons to **Compile** and **Apply** the script. There is also the new **Versioning** button, which helps you maintain versions of the script in addition to any maintained through source control. We will learn about this versioning system in an upcoming chapter.

- **Script Details**

  This panel will show and let you configure the properties of the script. One of the most important properties here is **Module Usage Bitmask,** which lets you specify which kind of scripts can reference the module. Then, there is **Provided Dependencies**, which lets you specify whether the module acts as a dependency for other modules, and **Required Dependencies**, which lets you specify other modules that this module may require. The **Library Visibility** option lets you make the module available in the right-click menu in the **Node** graph.

  If any input or output parameters are required for the script, you can specify them here. Other options in the **Script Details** panel let you specify **Description**, **Keywords** and **Script Metadata**.

Figure 7.9: The Script Details panel

- The **Parameters** panel

  The **Parameters** panel lists all the parameters that the module uses. You can add parameters to this panel by clicking on the + icon next to each category. Similar to the parameters in the main Niagara editor, we have different parameter categories here. The different parameter categories are as follows:

  - **System Attributes**: Attributes written in the System stage. Persistent and readable from anywhere.

  - **Emitter Attributes**: Attributes written in the Emitter stage. Readable from Emitter and Particle stages.

- **Particle Attributes**: Attributes written in the Particle stage. Readable from Particle stages.

- **Module Inputs**: Expose Module input to the **System** and **Emitter** editor.

- **Static Switch Inputs**: Set in Editor-only values.

- **Module Locals**: Non-persistent transient values used internally in the module.

- **Engine Provided**: Engine-provided read-only values.

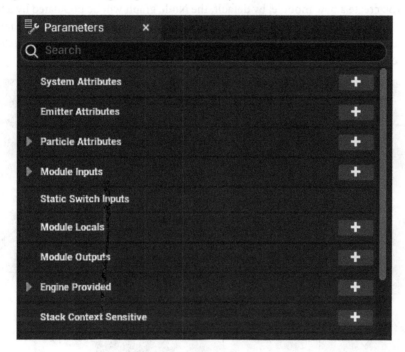

Figure 7.10: The Parameters panel

- The **Stats** panel

    This panel shows information such as the operations count for the module being edited. *Figure 7.11*, for example, shows us that the last operation count (**LastOpCount**) for our module is 7. The lower this value, the more efficient our module is.

Figure 7.11: The Stats panel

- The **Node** graph

  This is the main area of the editor in which you can edit the module script. It is visually similar to blueprints in Unreal, with similar navigation and operation methods. You can right-click anywhere to open a pop-up menu in which all the available nodes can be chosen from and dragged into the **Node** graph. Similar to blueprints, you can drag the white execution nodes and the colored data nodes.

  When you create a new module, by default, the **Node** graph will be populated by a template graph with the **Input Map**, **Map Get**, **Map Set**, and **Output Module** nodes. You will most likely keep them and add additional nodes to the graph. However, you can choose to delete these template nodes and create the **Node** graph from scratch.

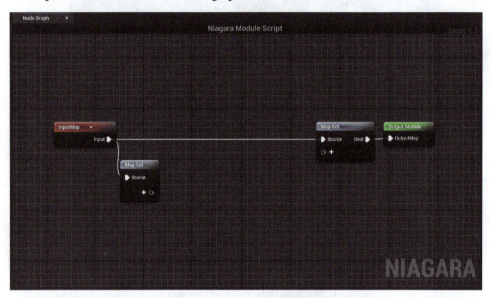

Figure 7.12: The Node graph

- The **Niagara Log** panel

  This is where the editor prints out the output of your compiled script. Any warnings or errors during compile time will show up here.

Figure 7.13: The Niagara Log panel

- The **Selected Details** panel

    This panel is similar to the **Details** panel in other Unreal editors. It shows the various properties of the node selected in the **Node** graph in detail.

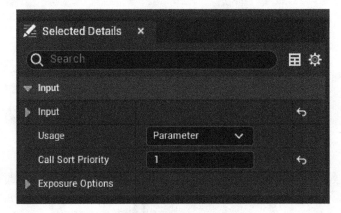

Figure 7.14: The Selected Details panel

We do not need to be familiar with every aspect of the editor to create basic modules, so do not stress over learning about everything in detail.

We will develop all our custom modules using this editor interface. Let us start developing one next.

Editing a Niagara Module to create custom effects

Now that we are familiar with the Niagara Module Script Editor, let us start working on the **Presence Detector** module we have created.

Before we start working on the actual code, let us first understand how the module works by looking at a simple example.

Understanding how the module works

This is an example where the module affects the size of a sprite based on the input values given by the user in the module's **Selection** panel.

In the **Node Graph** panel, a template graph node will have been created with the **InputMap**, **Map Get**, **Map Set** and **Output Module** nodes added. We will keep the template graph node and start with our Module script by clicking on the + sign on the **Map Get** node. We want the user to be able to specify the size of the particle. We will use uniform scaling for the particle size. This means we can use a **float** input instead of a vector. The new **float** input value pin will be in **Input Namespace**, as indicated by the **INPUT** label (see *Figure 7.16*):

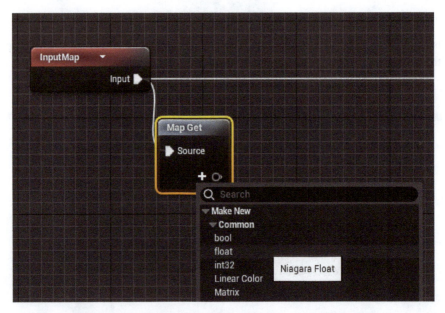

Figure 7.15: Adding a float Input value to the Map Get node

We will rename this pin `SizeOfParticle` by right-clicking on the pin and choosing the **Rename pin** option. This name will show up in the **Selection** panel when we add this module to the **Emitter** node.

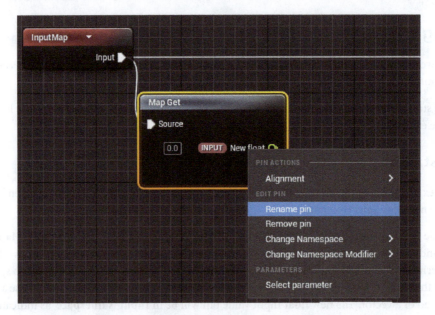

Figure 7.16: Renaming the Input pin SizeOfParticle

We want this **Input** value pin that we have added to control the size of the particle sprite. For this, we will have to add the **SpriteSize** parameter from the **PARTICLES** namespace to the **Map Set** node. We can do this by clicking the + sign on the **Map Set** node.

Figure 7.17: Adding the PARTICLES.SpriteSize parameter to the Map Set node

For our module to have an effect, we will need to connect the **SizeOfParticle** pin on the **Map Get** node to the **SpriteSize** pin on the **Map Set** node. This can be done by dragging a connector from the **SizeOfParticle** pin to the **SpriteSize** pin.

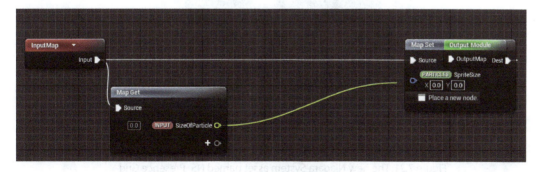

Figure 7.18: Connecting INPUT.SizeOfParticle to PARTICLES.SpriteSize

Since the **SpriteSize** pin is expecting a **Vector 2D** value and **SizeOfParticle** is a **float** value, the editor will automatically add a **float** value to the **Vector 2D** node, as shown in *Figure 7.19*.

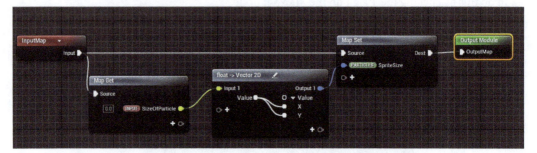

Figure 7.19: The float -> Vector 2D node automatically gets added

Our module is now ready to be used in an **Emitter** or **System** node.

Let us create the Niagara System to which we will add this module. We will use the **Empty** template.

Figure 7.20: Create a new Niagara System with an Empty template

Here, we will rename the system NS_PresenceGrid.

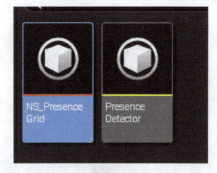

Figure 7.21: The new Niagara System asset named NS_Presence Grid

Now that the Niagara System has been created and named, open the system for editing, and from **Content Browser**, drag the **Presence Detector** module onto the **Emitter** node. The **Presence Detector** module will show up in the **Emitter** node.

Figure 7.22: Dragging the Presence Detector module into the Emitter node in the Particle Update section

On dragging the **Presence Detector** module into the **Emitter** node, the node should look as shown in *Figure 7.23*.

Figure 7.23: The Presence Detector module in the Emitter node

Select the **Presence Detector** module in the **Emitter** node. This will reveal the input parameters we specified in the **Map Get** node in the module. We added the pin in **Map Get**. This results in the parameter being shown in the **Selection** panel, as shown in *Figure 7.24*. The user will be able to modify this parameter or add Dynamic Inputs to it. We have therefore understood how the **Map Get** and **Map Set** nodes in the custom module work, allowing us to add custom properties to the module's **Selection** panel.

Figure 7.24: The SizeOfParticle property shows up in the Selection panel

Creating a Presence Detector particle effect

Now that we have understood how we can add input and output parameters to a custom module, let us get to work to create the functionality we want.

Let us clean up the module and delete the **SizeOfParticle** pin. We will add new pins that we require for our desired effect.

The first pin we will add is the **Position** module from the **PARTICLES** namespace, as seen in *Figure 7.25*.

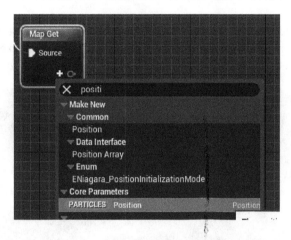

Figure 7.25: Adding the PARTICLES.Position pin to Map Get

This will let us read the position of the particles created by the emitter.

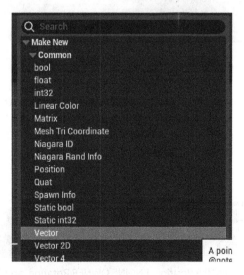

Figure 7.26: Adding a Vector parameter to Map Get

Next, let us add a **Vector** parameter in the **INPUT** namespace and rename it `PlayerPosition`. This will be the second pin on the **Map Get** node placed after the **PARTICLES.Position** pin.

Figure 7.27: Renaming the vector value to PlayerPosition

For ease of use, we would rather prefer the **PlayerPosition** pin to be the first pin in our node. To rearrange the order of the pins, you can right-click on the pins and move them up or down by choosing the appropriate option in the right-click pop-up menu, as you can see in *Figure 7.28*.

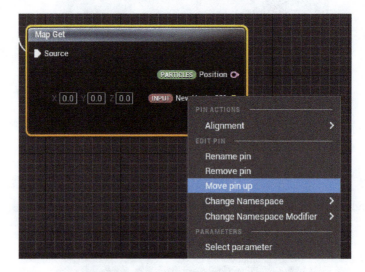

Figure 7.28: Moving the PlayerPosition pin up

We now need to read three more properties of the particle system owner, viz. **Position**, **Rotation**, and **Scale**. To be able to read these properties. we add three more pins to the **Map Get** node. Those are the **ENGINE OWNER Position**, **ENGINE OWNER Rotation**, and **ENGINE OWNER Scale**. These are read-only values that come from the engine itself. They can tell us the properties of the actor that owns the particle system. In this case, we are asking for the position, rotation, and scale of the actor owning this particle system.

Figure 7.29: Adding ENGINE OWNER Position, ENGINE OWNER Rotation, and ENGINE OWNER Scale

Now you might wonder why we collected the **ENGINE OWNER** properties. **PlayerPosition** will be in world space while our particle position will be in local space of the particle system. If the particles are going to interact with **PlayerPosition**, we need to apply a local transform to the particles. This can be done by adding an **ApplyLocalTransform** node. This node needs the transform values of the actor that owns the particle system. This is why we collected the **ENGINE OWNER** properties. Connect **PARTICLES.Position** to the **InputVector** pin. Niagara will automatically add **Position** to the **Vector** transformation node. Similarly, connect the **ENGINE OWNER Position**, **ENGINE OWNER Rotation**, and **ENGINE OWNER Scale** nodes to the **Translate**, **Scale**, and **Rotate** pins of the **ApplyLocalTransform** node respectively.

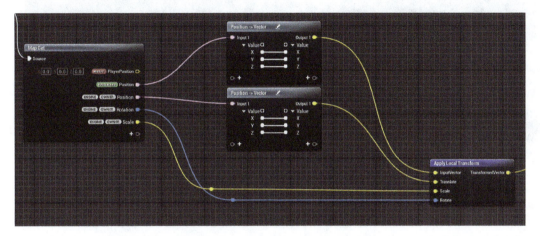

Figure 7.30: Adding the Apply Local Transform node and making the connections

We will connect **TransformedVector** from the **Apply Local Transform** node to the **Subtract** node to subtract it from **PlayerPosition**. To do this, connect pin **A** of the **Subtract** node to the **PlayerPosition** pin on the **Map Get** node and the **Transformed Vector** pin from the **Apply Local Transform** node to the **B** pin of the **Subtract** node. This will give us a vector relative to the player's position instead of relative to the world origin.

For a quick refresher on how vectors work, refer to the explanation in the *Vector mathematics and matrices and their representation in Niagara* section of *Chapter 2*. We are not interested in the direction specified by the vector and only want the magnitude, which we can get by connecting the subtraction result to a **Length** node. This gives us the distance of each particle from the player position. In effect, we can use this as an influence range to control the particles. Only the particles inside the influence range will be affected by whatever behavior we define in our module. To limit this range of influence, we will add a **Clamp** node. Let us set the **Max** value to 700.0. This will ensure that only particles at a distance of 700 units from the player position will be affected. We can tweak this value as per the creative requirements.

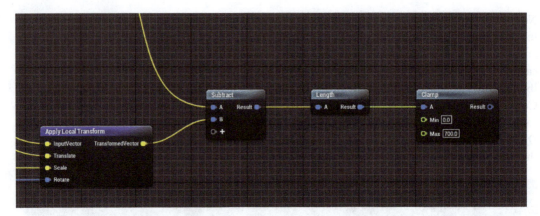

Figure 7.31: Adding Subtract, Length, and the Clamp node

We will add a **Multiply** node to enable us to scale the effect strength. Although optional, it is a good idea to have a multiplier to make the effect tweakable for creative control.

We need to convert the length value we have derived back into a Vector2D, as the effect of scaling of particles will act on the ground plane, which in the Unreal coordinate system is the XY axis. To do this, we first convert the clamped **Length** value to a 3D vector by using the **Make Vector** node. Since the input in our case is just a float, we will connect it to the **X** input pin of the **Make Vector** float. We will connect the **X** input to the **X**, **Y**, and **Z** pins of the output. (See the **Make Vector** node in *Figure 7.32a*). We can do this by dragging the connecting line from **X** to the appropriate pins in the **Make Vector** node. Normally, they are connected from *X to X*, *Y to Y*, and *Z to Z*. Once the value is converted into a 3dVector, we will convert it into **Vector 2D** by using the **Vector -> Vector 2D** node.

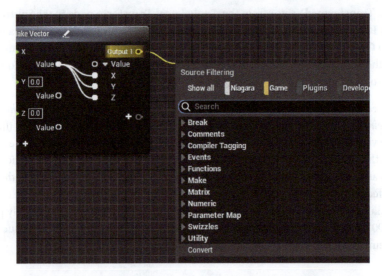

Figure 7.32a: Adding the Vector -> Vector 2D node by choosing Convert

To add the **Vector -> Vector 2D** node, drag a line from the **Make Vector** node's **Output 1** pin and select **Convert** from the pop-up menu. This will add a **Convert** node to the graph. Click on the + sign next to the left wildcard pin to change it to a **Vector** type. Similarly, change the right pin to a **Vector 2D** type. Rename the left pin `Input1` and the right pin `Output1` by right-clicking on the pin name and typing the new name in the text box under the **Edit Pin** section.

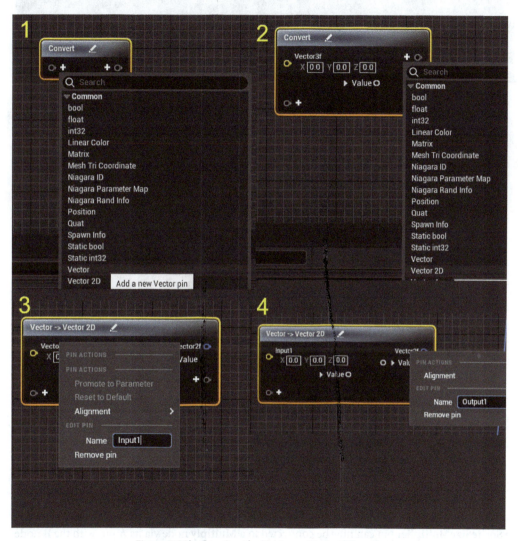

Figure 7.32b: Creating the Vector -> Vector 2D node

This node outputs only the X and Y values, while the Z value is eliminated. The output can now be connected to the **PARTICLES.SpriteSize** pin in our **Map Set** node, but we will give ourselves the ability to be more flexible creatively by adding a **Multiplier** node for **SpriteSize**.

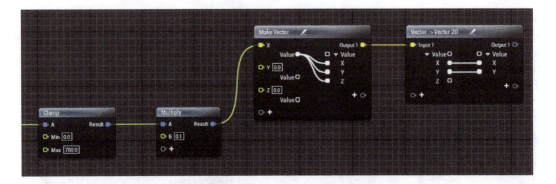

Fig. 7.32c: The whole conversion chain. Note that the Make
Vector has the input X connected to Output XYZ

To do this, we will get the multiplier value from the user by adding an input parameter to **Map Get**.

Let us add a **float** value input pin and we will name it `SpriteSizeMultiplier`. You can move the pin to a different position on the **Map Get** node if you want.

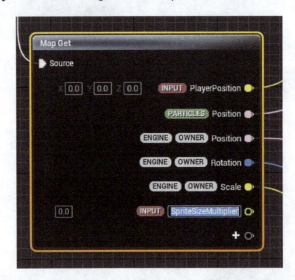

Figure 7.33: Adding the SpriteSizeMultiplier float pin to Map Get

The **SpriteSizeMultiplier** pin can now be connected to a **Multiply** node via its **A** pin with the **B** node of the **Multiply** node connected to the **Vector2D** value we evaluated earlier. We will connect the value of this **Multiply** node to the **PARTICLES.SpriteSize** pin in **Map Set**. We are now ready to test our custom module. We already added the module to the **Emitter** node, which shows up as **Presence Detector** under **Particle Update**.

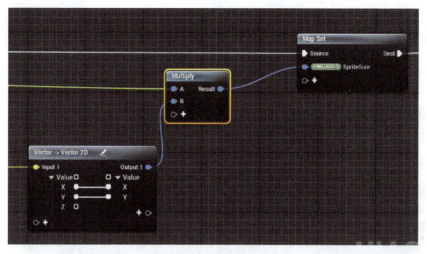

Figure 7.34: Adding the Multiply node and connecting its output
to the PARTICLES.SpriteSize pin on Map Set

We now need to create a grid of particles lying on the ground. Let us add the **Spawn Particles in Grid** module to **Emitter Update**. Niagara will mark it with a red dot, indicating that there are some issues with this module. You will get additional information about the issue in the **Selection** panel. In this case, we are informed that there are unmet dependencies for the module to work.

Figure 7.35: A warning red dot next to the Spawn Particles in Grid module

Unmet dependencies usually mean that Niagara needs additional nodes to be added for the node to work. Now, we might not always be aware of what the additional nodes are. Thankfully, if you select the module in the **Emitter** node, the **Selection** panel will show what modules are needed. In this case, Niagara suggests that we add the **GridLocation** module. A **Fix issue** button also pops up and it is recommended that you press that button rather than trying to add the dependency module manually.

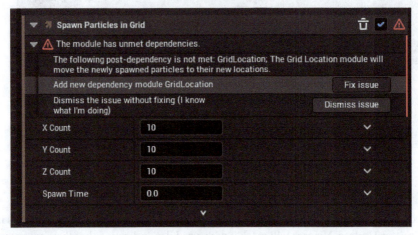

Figure 7.36: Niagara asking us to add the Grid Location module and offering the Fix Issue option

Once you click on the **Fix issue** button, the **Grid Location** module is added in the **Particle Spawn** section. The red error dot also disappears, and we can move to the next step.

Figure 7.37: The red dot has disappeared and the Grid Location module has been added

In the **Preview** window, you will see that the particles are emitted in a square grid shape.

Figure 7.38: The Spawn Particles in Grid module created a bunch of particles arranged in a 3D grid

We need to spread the particles out. We can do that by changing **X** and **Y** to 100.0 in the **XYZ Dimensions** parameter in the **Grid Location** module. We will keep the **Z** value as 1.0 as we want the grid to be spread only on the ground.

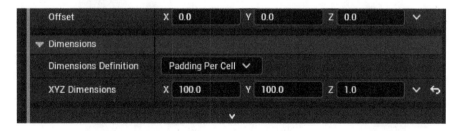

Figure 7.39: Spreading the particles by setting the grid spacing to 100 in X and Y

In the **Spawn Particles in Grid** section, we will change **X Count**, **Y Count**, and **Z Count** all to 50. This can be tweaked as per your creative requirements. **Z** can be kept as 1 if you want.

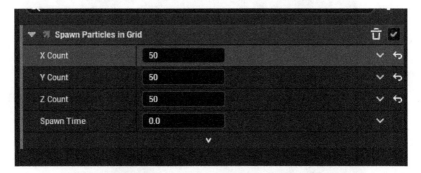

Figure 7.40: Setting the number of grid lines by defining the X Count,
Y Count, and Z Count properties for Spawn Particles in Grid

Check the **Preview** window and you should see something similar to *Figure 7.41*. We should have a 50x50 grid of particles.

Figure 7.41: The 50x50 grid that we just created

If you do not see anything in the **Preview** window, check the **SpriteSizeMultiplier** value by clicking on the **Presence Detector** module and looking at the value of **SpriteSizeMultiplier** in the **Selection** panel. If it is **0**, then we should change it to **1**. A better idea would be to set the default to **1**. This will prevent other users from being confused when no particles show up because of the particles being set to a scale of **0** when they use the custom module.

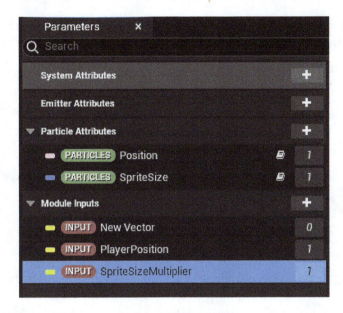

Figure 7.42: Selecting the SpriteSizeMultiplier parameter in the Parameters panel

To set default value of the **SpriteSizeMultiplier** to 1.0, select it in the **Parameters** panel. Now, check the **Selected Details** panel, and, under **Default Value**, change the value to 1.0.

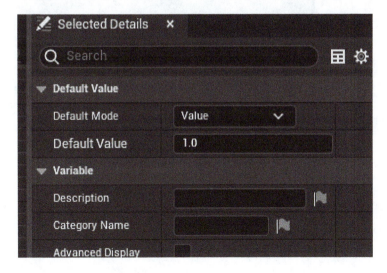

Figure 7.43: Changing Default Value to 1 for SpriteSizeMultiplier

Now, select our custom module, **Presence Detector**, in the **Emitter** node. Set **PlayerPosition** to 0.0, 0.0, and 0.0 and **SpriteSizeMultiplier** to 1.0. Ensure that the **Presence Detector** module is active.

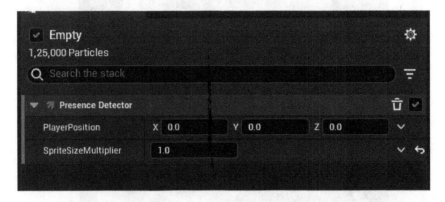

Figure 7.44: Setting the default PlayerPosition and SpriteSizeMultiplier values to test the functionality

The blue checkmark next to the module name should be visible, which indicates that the module is active.

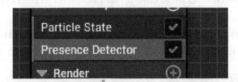

Figure 7.45: The Presence Detector module is active

If everything has been done correctly, the particles should display the behavior shown in *Figure 7.46*. There should be an area at the center of the particle system where the particles are of a smaller size than the rest of the grid. You can play with **PlayerPosition**'s **X** and **Y** values to move the area of smaller-sized sprites around in the grid.

Figure 7.46: Behavior of the particle system with the Presence Detector module active

For our particle system to work in a level, where the area of smaller sized sprites follow the player's movements, we will need to read the position of the player character in the level and pass those coordinates to the **PlayerPosition** property in our **Presence Detector** module.

We will do this by creating a custom parameter in the **User** namespace and then using a blueprint to read the player character position and passing it to this custom parameter. This custom parameter will be connected to the **PlayerPosition** parameter in our module.

To create the custom parameter in the **User** namespace, click on the + sign in the **User Exposed** section. Since we are reading the player position, we will create a value of type vector.

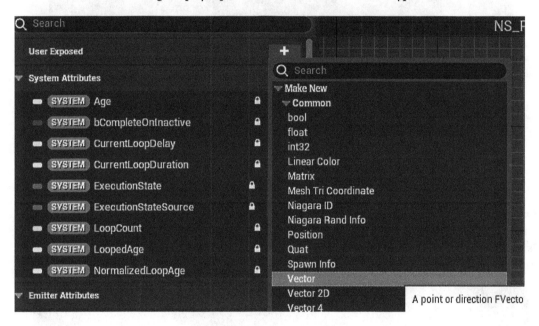

Figure 7.47: Creating a USER.PlayerPosition vector value in the User Exposed section

This will add a **Vector** parameter under **User Exposed**. Let us rename this `PlayerPosition`. Even though we have chosen the same name as we chose in our **Presence Detector** module's **Map Get** node, we need not worry as this parameter is in the **User** namespace.

Figure 7.48: The USER PlayerPosition vector value created

We are going to connect this newly created **USER.PlayerPosition** parameter to the **PlayerPosition** parameter in our **Presence Detector** module. To do this, drag the **USER.PlayerPosition** parameter and drop it into the **PlayerPosition** parameter. A blue dotted line highlight will appear around the eligible fields as you drag the **USER.PlayerPosition** parameters toward them.

Figure 7.49: Connecting the PlayerPosition parameter in the
Presence Detector module to USER.PlayerPosition

Once **USER.PlayerPosition** is dropped into **PlayerPosition**, the **USER.PlayerPosition** label with a chain link will show up instead of input boxes. This means that the module's **PlayerPosition** parameter is now driven by the **USER.PlayerPosition** parameter. We will drive **USER.PlayerPosition** using a public variable later in this chapter when we create a blueprint actor containing our Niagara System.

Figure 7.50: The chain link icon indicating that PlayerPosition is being driven by USER.PlayerPosition

Let us test whether these steps are working as expected. For that, if we change the values of **USER. PlayerPosition**, then the **PlayerPosition** should change and affect the behavior of the particle system. To change **USER.PlayerPosition**, in the **System** node of the **NS_PresenceGrid** Niagara System, select **User Parameters**.

Figure 7.51: The User Parameters module in the NS_PresenceGrid System node

In the **Selection** panel, you should see the **USER.PlayerPosition** parameter displayed. We know that this is driving the **INPUT.PlayerPosition** parameter.

Figure 7.52: Changing the USER.PlayerPosition values in the User Parameters module

Change **X** and **Y** values of the **USER PlayerPosition** parameter and you should see the area with the smaller sprites move around in the particle grid in the **Preview** panel.

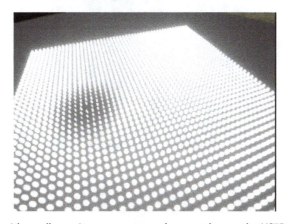

Figure 7.53: The area with smaller sprites moves around as you change the USER.PlayerPosition values

We want to take this a little further and add more features to this particle system. We want the particles in the grid to have a certain color while the color of the particles that are near the character should change. Let the particles be red in color and become white when the player character approaches them. Essentially, this means that the particles that scale down as the player approaches them will also be the ones changing color.

To achieve this effect, we will have to modify our custom module, **Presence Detector**.

We will add an additional parameter to the **Map Set** node in our module.

Figure 7.54: Adding a PARTICLES.Color parameter to Map Set

We want to set the particle color, so this parameter will be of type **PARTICLES.Color**.

Figure 7.55: The Map Set node after adding a PARTICLES.Color parameter

From the **Clamp** node, we will create a branch. Since we have clamped the value from **0** to **700**, we can use a **Remap Range** node to map the values from the **0** to **700** range to the 0 to 1 range. This will help us define the RGB values of the linear color, which we will then feed into the **PARTICLES.Color** parameter that we added to the **Map Get** node.

Since we want the particles to be red in normal circumstances, we will feed the value from the **Remap Range** into the **R** pin of a **Make Linear Color** node. Now, in normal cases, when the player character is away from the particle, the value coming out from the **Remap Range** node is going to be closer to 1.0. This will mean that the **Make Linear Color** node will output a reddish color. We will take the value coming out of the **Remap Range** node and subtract it from 1.0 using a **Subtract** node. As a result, when the player character approaches the particle, the value output from the **Subtract** node will lean toward 1.0.

This will make the particle color become white with a cyan hue, thus achieving our need for a color change from red to white.

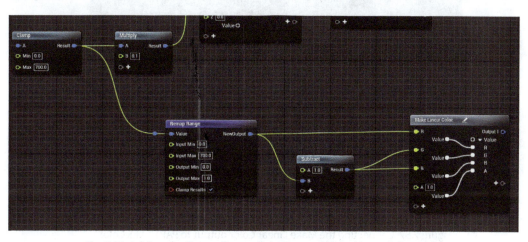

Fig. 7.56: Adding the Remap Range, Subtract, and Make Linear Color nodes

Let us connect the output from the **Make Linear Color** node to the **PARTICLES.Color** pin in **Map Set** that we added earlier. Our custom **Niagara Module Script** asset is now ready.

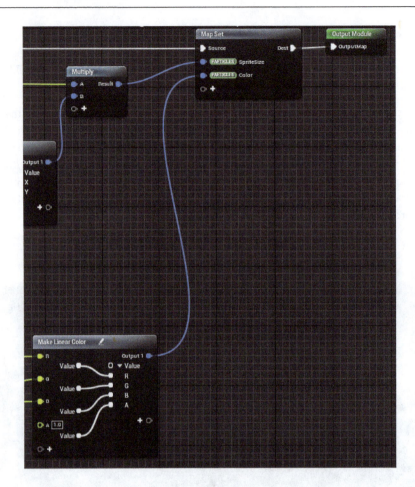

Figure 7.57: Connecting the Make Linear Color output to PARTICLES.Color

The full script should look like *Figure 7.58*:

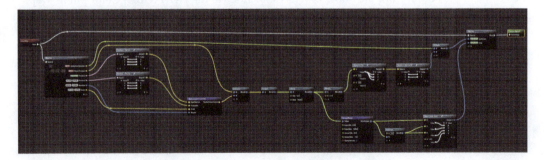

Figure 7.58: The full Presence Detector module script

The particle system in the **Preview** window should look like the following figure, with the particles having a reddish hue overall and changing to white at the center of the grid with **PlayerPosition** set to 0.0, 0.0, and 0.0. Let us now start creating the blueprint, which will read the position of the player character in the game, in our case the third-person character, and pass the information to the Niagara particle system.

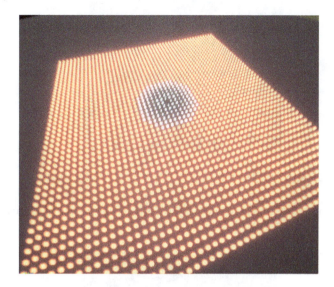

Figure 7.59: The resultant behavior of our Presence Detector module

Create a blueprint class containing a Niagara System

Now that our **Presence Detector** module is ready, we want **USER.PlayerPosition** to be driven by the position of the player in the game. For this, the player position in the game will have to be read every frame and passed on to the Niagara particle system. The particles will then change in response to the player's movement.

Let us create a new blueprint of the **Actor** class and let us name it BP_Presence Detector.

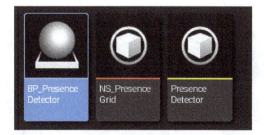

Figure 7.60: Creating a BP_PresenceDetector blueprint of class Actor

Open the blueprint in the **Blueprint** editor by double-clicking on it. In the **Components** panel of the **Blueprint** editor, click on the + **Add** button to reveal the **Components** menu. Choose **Niagara Particle System Component**.

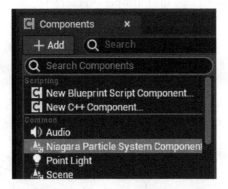

Figure 7.61: Adding Niagara Particle System Component

This will add a Niagara particle system component to the blueprint, which will be named **Niagara** by default. Feel free to rename it if needed.

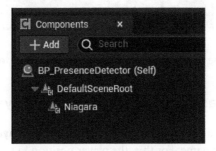

Figure 7.62: The Niagara particle system component added and named Niagara by default

Now, select the **Niagara** component. In the **Details** panel of the blueprint, go to the **Niagara** panel, and set **Niagara System Asset** to **NS_PresenceGrid**. This is the same Niagara System we created, which contains our custom module.

Figure 7.63: Setting Niagara System Asset to NS_PresenceGrid

In the **Event** graph of our blueprint, we will add the functionality to read the position of the player character. We will add this functionality to **Event Tick** to enable us to read and update the position of the player for every frame. In the interest of performance, this functionality may also be added using the **Set Timer By Event** node with a slightly smaller update frequency.

We will read the position of the character by using the **Get Player Character** node and then read the location by using the **Get World Location** node. It is recommended you choose the **Capsule** Component to read the location, but depending on the character you have, you are free to choose the appropriate component.

Let us drag the **Niagara** component into the graph. From the **Niagara** component, we can drag out and add the **Set Niagara Variable (Vector3)** node. There will be various options for **Set Niagara Variable**, such as **float**, **linear color**, etc. We chose **Vector3** because the variable we will set is **USER. Player Position**, which is **Vector3**.

Your **Node** graph should look something like this:

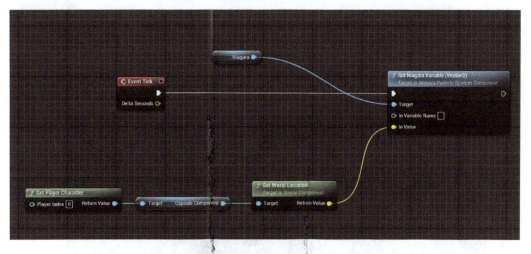

Figure 7.64: The blueprint script in BP_PresenceDetector

Now, go to the **NS_PresenceGrid** Niagara System tab and right-click on the **USER.Player Position** parameter in the **Parameters** panel. Copy the reference to that parameter.

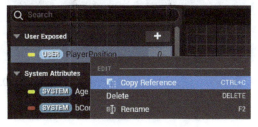

Figure 7.65: Copying the reference from USER.PlayerPosition

Now, come back to the blueprint graph and in the **Set Niagara Variable (Vector3)** node, paste the copied reference into the **In Variable Name** slot. The pasted text should say **User.PlayerPosition**. The **User** namespace is important. We are now ready to test our blueprint.

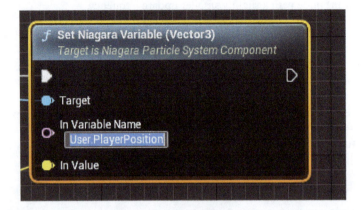

Figure 7.66: Pasting the reference copied in Figure 7.65

Drag the blueprint into the scene. To keep things simple, we haven't added calculations for the relative positioning of the particle system. So, we will need to ensure that the particle system is at the origin, i.e., the (0.0, 0.0, and 0.0) location. Set the **Transform** values of the blueprints, as shown in *Figure 7.67*.

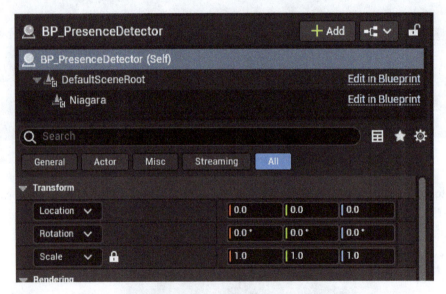

Figure 7.67: Setting Location and Rotation of BP_PresenceDetector to (0.0, 0.0, and 0.0)

Our blueprint containing our Niagara System, and our custom module in it, should now be working. To test it, press the **Play** button on the toolbar and move the player character around.

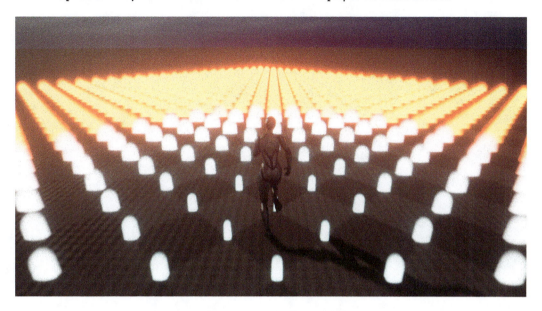

Figure 7.68: As the player runs around, the particles near the player
become smaller and change from red to white

You should see the particle system react to the presence of the player by scaling down and changing color as the player character approaches them.

Summary

Learning how to create custom Niagara modules in the Niagara framework helps us extend the functionality of the framework to suit our needs better. Niagara modules are the building blocks of Niagara Systems, and by creating custom modules, we can tailor the systems to specific tasks or scenarios.

In this chapter, we learned how to create these custom module assets and encapsulate complex functionality in a single module. In the next chapter, we will see a different way in which custom modules can be implemented locally in a system.

8

Local Modules and Versioning

Now that we are familiar with custom modules, we will learn about a different type of custom module called **Local Modules**. In this chapter, we will learn how to create these **Local Modules** (previously known as **Scratch Pad** modules). **Local Modules** are embedded in the particle system and do not exist as independent reusable assets (although you can copy and paste them and convert them to custom modules). We will also be introduced to a versioning system, which lets us manage different versions of the custom modules that we will create.

The topics we will cover in this chapter are as follows:

- Exploring **Local Modules**
- Publishing modules and versioning

Technical requirements

You can find the project we worked on in this book here:

```
https://github.com/PacktPublishing/Build-Stunning-Real-time-VFX-
with-Unreal-Engine-5
```

Exploring Local Modules

In the previous chapter, we learned how to create custom modules to achieve behaviors that cannot be achieved using existing Niagara modules. Custom modules can be reused in other particle systems if we need to. But there are many occasions where we want a one-off custom module and we do not intend to reuse it in other particle systems.

For such situations, we should create a local module. **Local Modules** are embedded in the particle system and cannot be reused in other particle systems. **Local Modules** are also useful for testing out and prototyping new modules.

Local Modules can be edited in their own editor, which is similar to the Niagara Module Editor.

In *Unreal Engine 5.1*, a new panel has been introduced to replace the **Scratch Pad** panel from *Unreal 5.0* and below. The new panel is called the **Local Modules** panel. The functionality remains the same as the earlier **Scratch Pad** panel with the exception that the module editor is now separate from the rest of the panels. The module editor now only appears when you double-click on a newly created Local Module to edit it.

Figure 8.1: The Local Modules tab panel grouped with the User Parameters and Parameters tabs

The **Local Module** panel is blank when you have no **Local Modules**. The panel will fill up with the relevant data as you start creating modules. You can start creating a new module by clicking on the + icon next to the **Modules** label. However, the other method of creating a Local Module is more common as it is similar to the standard process of adding a module.

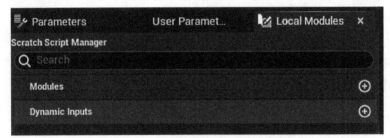

Figure 8.2: The Local Modules tab panel is empty until you create a Local Module

The other method of creating a Local Module is to click on the + icon in any of the various sections in the **Emitter** nodes in a particle system, such as **Particle Update** or **Particle Spawn**, to bring up the **Add New Module** pop-up menu. Scroll down to the bottom of the pop-up menu and select **New Scratch Pad Module**. **Local Modules** were called **Scratch Pad** modules in previous versions, and hopefully this menu option will be renamed **New Local Module** in a future version.

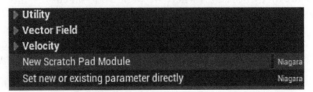

Figure 8.3: Adding a new Local Module

This will take you to the **Module Editor** with a new **ScratchModule** that's ready to be renamed.

A template node graph is also created for you to edit. The node graph has four important nodes. The **InputMap** and **Output Module** nodes receive and output the particle state from the **Overview** node stack. The **Map Get** node is where you can add parameters that you want to read from the particle system or properties you want to input from the user.

The **Map Set** node will set particle system parameters that may eventually affect particle behavior or be used by other modules to ultimately act on the particles.

Figure 8.4: The Local Module editor

Let's rename our new Local Module `LightningSparks`. You can choose any name, but of course, it is best to have a descriptive name.

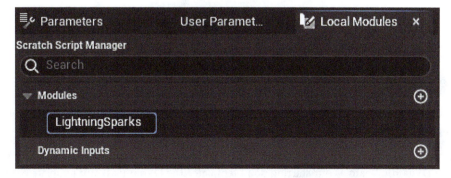

Figure 8.5: Naming our Local Module LightningSparks

The Local Module will be added to the **Emitter** node. A paper and pen icon on the right of the label indicates that it is a Scratchpad (Local) module. Your emitter module should look like *Figure 8.6*.

Figure 8.6: Our Local Module showing up in the Emitter node

Create a lighting effect using modules

Let's create a particle system where a lightning bolt strikes the ground at random locations from the sky. The lightning bolt will be red, and the bottom part of the lightning bolt will be blue in color.

To do this, let's start by creating a new Niagara particle system. This time, instead of the **Fountain**, let's choose **Dynamic Beam** as our starting point template. The **Dynamic Beam** template has most of the stuff that we need for this effect already set up.

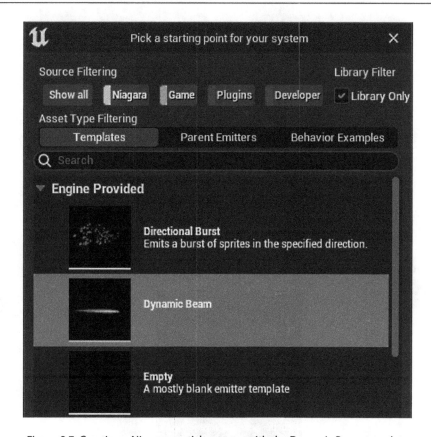

Figure 8.7: Creating a Niagara particle system with the Dynamic Beam template

Let's name our new particle system `NS_ColouredLightning`.

Figure 8.8: Naming our particle system NS_ColouredLightning

Double-click on **NS_ColouredLightning** to open the Niagara Editor. We should see the **Dynamic Beam** emitter that comes with the template. When compared to the **Fountain** emitter, you will notice this emitter has a different set of modules. There are a bunch of beam-related modules, and instead of a **Sprite** renderer, it has a **Ribbon** emitter.

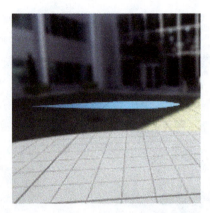

Figure 8.9: The particle system appearing in the preview as a beam

In the **Preview** panel, you should see a straight horizontal beam.

Figure 8.10: The overview node for the Dynamic Beam. Note the
Ribbon Renderer module in the Render group

We will modify this emitter to achieve the behavior that we want. Let's start with the Emitter State module.

In the **Emitter State** module, set **Inactive Response** to **Kill (Emitter and Particles Die Immediately)**. In **Loop Duration**, add a **Random Range Float** dynamic input, and in the input parameters that appear, set **Minimum** to 0.1 and **Maximum** to 0.2.

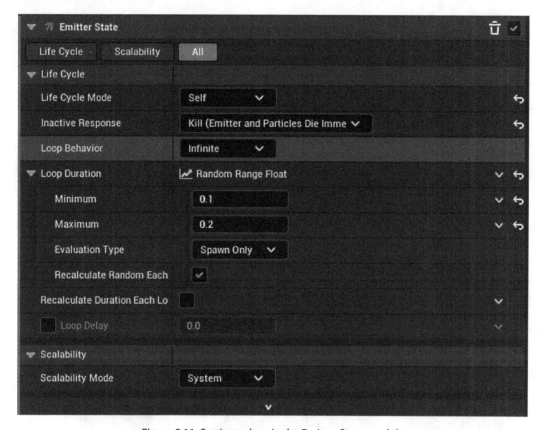

Figure 8.11: Setting values in the Emitter State module

Let's focus on the **Beam Emitter Setup** module. We are going to make some major changes here. The module should currently look like this:

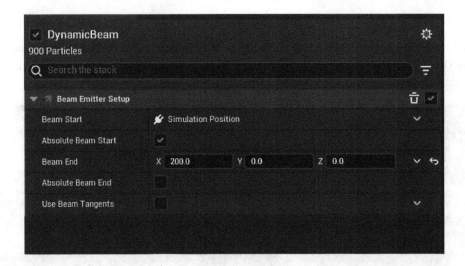

Figure 8.12: The Beam Emitter Setup module before we start modifying it

Beam Start defines the position at which the lightning bolt will start. Technically, our lightning bolt will start from the base and end in the sky. We want our lightning bolt to look as if it is striking random places on the ground and starting from one place in the sky. Let's start by adding a **Random Range Vector Dynamic** input to **Beam Start**. We need to define the area where the lightning bolt will strike. Let's define a rectangular area that's 400 x 800. To achieve that, in **Minimum**, enter the values -200 for **X**, -400 for **Y**, and 0 for **Z**. In **Maximum**, enter the values 200 for **X**, 400 for **Y**, and 1 for **Z**.

Let's define the top of the beam by entering the following values for **Beam End**: 0 for **X**, 0 for **Y**, and 625 for **Z**. The beam will start 625 units from the base of the particle system.

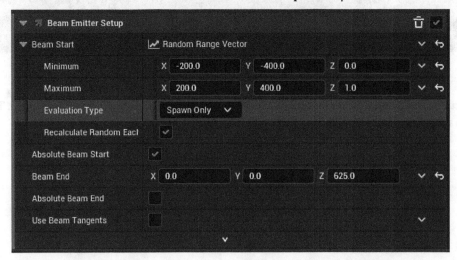

Figure 8.13: Adding Random Range Vector Dynamic Input to the Beam Start property

Our beam should behave as shown in *Figure 8.14*. It will jump to random start positions and will be straight, with a height of 625 units.

Figure 8.14: The beam jumps to random start positions while pointing upwards to a height of 625 units

We want our beam to be a bit curved as we would like it to represent an arcing lightning bolt. To do this, start by checking **Absolute Beam End** and **Use Beam Tangents**. This will update the panel to add more properties. A **Beam Start Tangent** property and a **Beam End Tangent** property will have been added to the panel.

For **Beam Start Tangent**, we will modify the property vector from **Engine.Owner.SystemXAxis** to a Local value, which we will choose by selecting **New Local Value** from the **Dynamic Input** menu.

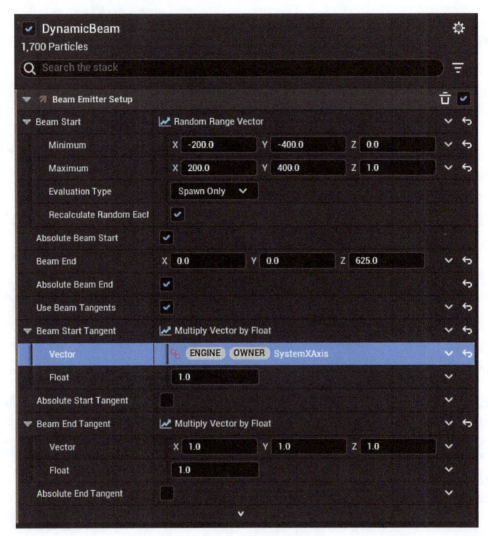

Figure 8.15: Checking the Use Beam Tangents check box and modifying
the Beam Start Tangent Vector value to a Local Value

For the **Vector** property, enter the values 0.4 for **X**, 0.5 for **Y**, and 0.6 for **Z**. For the **Beam End Tangent Vector** property, enter the values 1.0 for X, 1.0 for **Y**, and -0.2 for **Z**. Feel free to modify these values as you wish. Modifying these values will change the curvature of the beam because you are modifying the start and end tangent values of the beam.

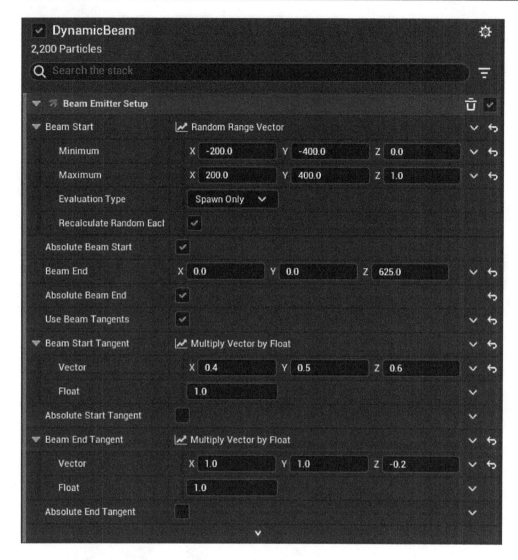

Figure 8.16: Setting the values of the Vector property in Beam Start Tangent and Beam End Tangent

The beam that we have at this point seems to stay in one place, which is not what we would expect a lightning bolt to do. We need it to be changing its shape rapidly as a lightning bolt would. So, let's reduce its lifetime. Let's change the **Lifetime** property in the **Initialize Particle** module to 0.02.

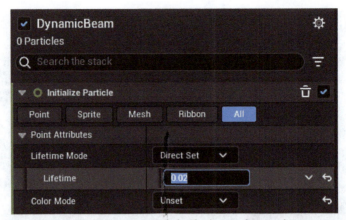

Figure 8.17: Setting Lifetime to 0.02 to make the beam change shape more rapidly

The beam width is very even, and the lightning bolt would look good if it had a random width. There is a **Beam Width** module, which has a **Float from Curve** dynamic input to shape the Dynamic Beam into a teardrop. Let's get rid of **Float from Curve** and replace it with **Random Range Float**.

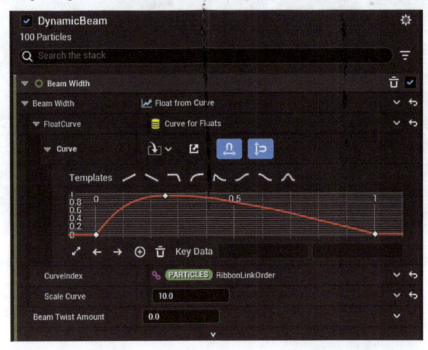

Figure 8.18: The Beam Width property as it is in the Dynamic Beam template.
We will change from Float from Curve to Random Range Float

Set the **Minimum** value of **Beam Width** to 5 and set **Maximum** to 20. For some extra variation, let's add a **Beam Twist Amount** of 90. The width of the lightning bolt will now be more random.

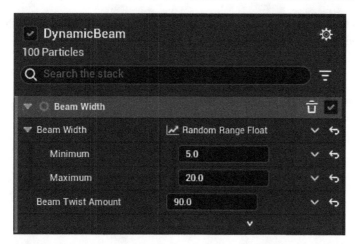

Figure 8.19: Minimum and Maximum values set for Beam Width. The Beam Twist Amount is also set to 90

Now, our lightning bolt needs to be shaped like a curved arc. To get it to look curved, change the **Curve Tension** value in the **Ribbon Renderer** module, under **Tessellation**, to 0.5. The Dynamic Beam that we started out with will now be a smooth curved beam striking at random places on the ground.

Figure 8.20: Adjusting the curvature of the beam with the Curve Tension property

We need the beam to be jagged to make it look like a lightning bolt. We can achieve this by adding a **Jitter Position** module in the **Particle Update** section.

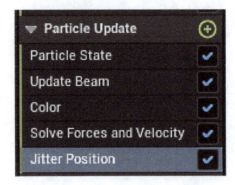

Figure 8.21: Adding the Jitter Position module

In the **Jitter Position** module, set the **Jitter Amount** to 10.

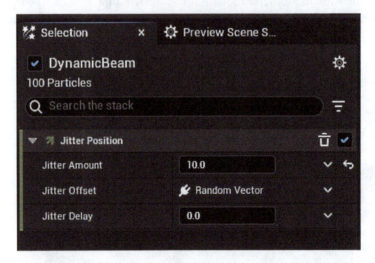

Figure 8.22: Setting the Jitter Amount in the Jitter Position module to 10.0

With the current setup, the bolt is a bit difficult to see in the preview window. The **Color** module from the **Dynamic Beam** template is reducing the intensity of the bolt a bit. Let's disable the **Color** module by unchecking it.

Figure 8.23: Disable the Color module

In the **Preview** window, we should now see a jagged lightning bolt hitting random points on the ground. The bolt will be white because we disabled the Color module.

Figure 8.24: The spark behavior as a result of all the modifications
we have done to the Dynamic Beam node

As per our brief at the beginning of this exercise, we would like the bolt to be glowing red, and the part of it that touches the ground should be glowing blue.

We cannot achieve this through existing modules. We will therefore need to write a custom module. However, instead of writing a new separate custom module, we can also write a Local Module that will be a part of only this particle system.

We will need to add this module to **Particle Update**.

Click on the + button next to the **Particle Update** label and choose **New Scratch Pad Module** from the menu.

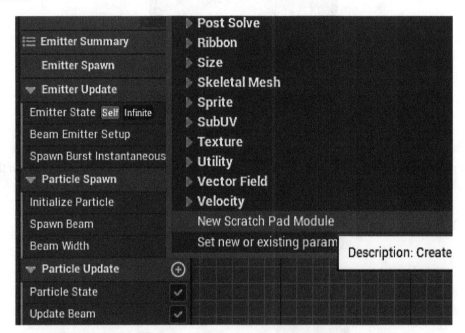

Figure 8.25: Adding a New Scratch Pad Module in the Particle Update group

We will be switched to the Local Module editor interface where we will rename the module to **ModifyColor** under the **Scratch Script Manager > Modules** section.

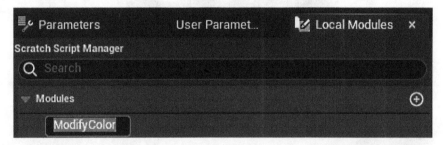

Figure 8.26: Naming our Local Module ModifyColor in the Scratch Script Manager

A template node graph will have been created for us. You will recall that this is the same workflow as when we created a separate custom Niagara module.

Figure 8.27: The ModifyColor Local Module looks very much like a
Custom module with the same nodes available at startup

The end of the beam that strikes the ground needs to be blue for a certain length. Remember that the beam is actually starting from the ground and moving to the end point 645 units in the Z direction.

A beam of light is made up of a series of particles that stretch from the start point to the end point of the beam. Each particle has a **Particle Index**, with the first particle having an index of 0. The index number increases along the length of the beam. We need to identify the particles with an index less than a certain index number and color them blue, while we leave the rest of the particles red.

Let's read the **Particle ID** in the **Map Get** node. Click on the + sign next to the gray pin on the **Map Get** node and choose **PARTICLES ID** from the menu that pops up.

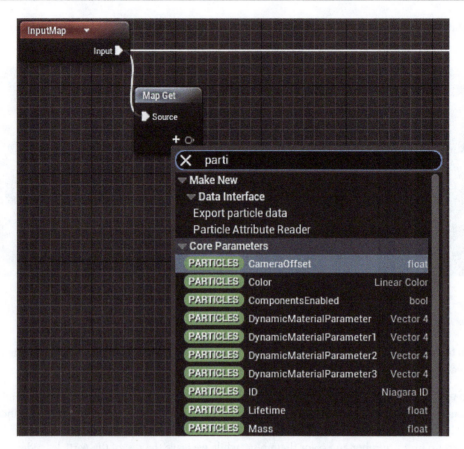

Figure 8.28: Adding the PARTICLES.ID parameter to Map Get

This ID is a struct that needs to be broken as we are interested only in the index. You can break the Particles ID struct by clicking on the + sign next to the gray pin in **Map Get** and adding a **Break Niagara ID** node from the pop-up menu. One of the output pins of the **Break Niagara ID** node will be the **Index** pin. Drag from the **Index** pin and add a **Greater Than** node. We will use the **Greater Than** node to find particles whose index is less than **10**. The **Greater Than** node returns a Boolean value. We will decide the color of the lightning bolt particles based on the value of the Boolean. If the value is true, we will make the particles red, and if the value is false, we will make the particles blue.

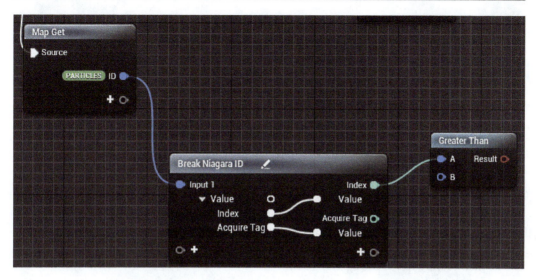

Figure 8.29: Adding the Break Niagara ID and the Greater Than nodes

We will have to compare the value of **A** (of the **Greater Than** node) with the value of **B** that we will set to 10. This way, we can find particles with an index less than or equal to 10. However, before we do that, we need to convert **B** to a 32-bit integer. We need to do this because without converting the value to a 32-bit integer, we won't get the appropriate slot in the node to enter the comparison value of 10. The conversion is done by right-clicking on the **B** pin and choosing **Convert Numeric To > int32**.

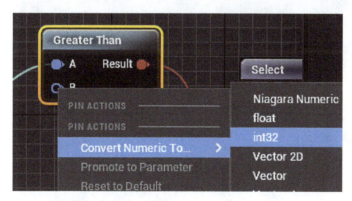

Figure 8.30: Converting the B pin from Numeric to int32 (32-bit integer)

B will be converted into an integer, which is evident because the color of the pin changes from blue to green. Let's enter a value of 10 for **B**. You can change this value to adjust the length of the blue part of the bolt.

Figure 8.31: Setting the value of the B pin to 10

Now, based on the Boolean result returned by the **Greater Than** node, we apply a color to the lightning bolt. The decision of which color to apply will be taken by the **Select/If** node, which is labeled as **Select** once it's added. To add the node, right-click and choose the **Select/If** option. Connect the **Result** pin from the **Greater Than** to the **Selector** pin in the **Select** node.

The **Select** node has **NiagaraWildcard** values for input and output. A **NiagaraWildcard** type lets you convert it to any type of Niagara data. In our case, we want the **NiagaraWildcard** to be **Linear Color**, which is what we will apply to the lightning bolt.

Figure 8.32: The Select node

To convert the wildcard values to **Linear Color**, click on the pencil icon on the left of the output pin labeled **NiagaraWildcard**.

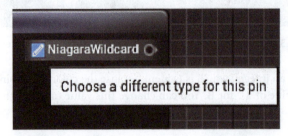

Figure 8.33: Click on the pencil icon the change the data type of the NiagaraWildcard pin

Choose **Linear Color** from the menu that pops up. All the pins should now convert to **Linear Color**.

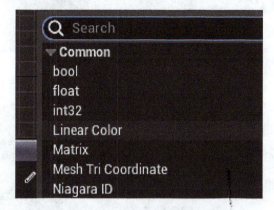

Figure 8.34: Choose Linear Color data type

The input **LinearColor** pins will have a default value set to the color white.

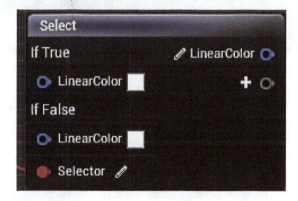

Figure 8.35: The Select node with the data type changed. Note the input pins data type also changes

There are two input pins here. The values relayed from the input pins to the output pin depend on the **Selector** Boolean value, that is, whether it is **True** or **False**. We need the particles to be red if the particle index is greater than 10 and blue if it is less than 10. We also need the color to glow. You can set the colors by clicking on the color swatches next to the input **Linear Color** nodes. Select **Red** for the **True Linear Color** input and **Blue** for the **False Linear Color** input. Set the **V** in HSV to 500 to make the color glow.

Figure 8.36: Setting the colors of the Linear Color pins in the Select node

The particle color of our system will depend on color emitted by the output pin of the **Select** node. As we saw earlier, if the input to the **Selector** pin of the **Select** node is true, then we will use **Red**, and if it is false, we will use **Blue**. This true/false will be determined by the **Greater Than** node. To apply the color output from the **Select** node, we need to add a **PARTICLES Color** pin to the **Map Set** node. Add that pin by clicking on the + icon on the **Map Set** node.

Figure 8.37: Adding the PARTICLES Color pin to the Map Set node

Connect the output of the **Select** node's **Linear Color** output pin to the **Map Set** node's **PARTICLES Color** pin.

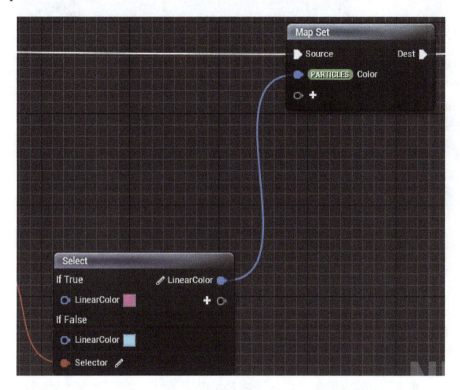

Figure 8.38: Connect the LinearColor output pin of the Select node
to the PARTICLES.Color pin of the Map Set node

We need to do a few things before the **Local Module** compiles correctly. Without that, you may see error messages.

In the **Emitter Properties** module, check the **Local Space** checkbox. This will ensure that the **Beam End** value moves with the Niagara emitter's location. Without **Local Space** being checked, the emitter beam end position will be locked to a specific world location, in this case to the world origin, independent of the Niagara emitter transform values. In other words, without it, the particle effect will end up at the world origin instead of the location where you placed the Niagara system.

You also need to check the **Requires Persistent IDs** checkbox. This is needed for our particle ID comparison in our Local Module using the **Greater Than** node to work.

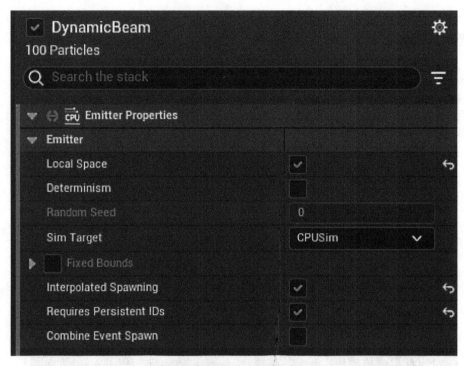

Figure 8.39: Setting the Local Space and the Requires Persistent IDs
checkboxes to on in the Emitter Properties module

We are done with our particle system and the Local Module. To see the Local Module's behavior being applied to the particle system, click the **Apply** button at the top of the module editor panel.

Figure 8.40: Click the Apply button for the Local Module to take effect

You should see the particle system behave as per our brief, with the lightning bolt being mostly red and blue at the end of the bolt. The bolt will also randomly move around.

Figure 8.41: The final result achieved with our Local Module

The **DynamicBeam** emitter will look like *Figure 8.41* if you have followed all the instructions. You may see a few modules have a red dot. This is a UI refresh bug; clicking the modules with the red dot should fix the issue.

Figure 8.42: The DynamicBeam emitter after all the changes

In this section, we learned about the Ribbon renderer as well as how to use a Local Module in practice.

As we develop modules and share them with the team, keeping track of different versions of the modules becomes important. In the next section, we will see how we manage this process.

Publishing modules and versioning

During development, particle assets are always being worked upon, even when they are being used in production. It may be necessary to release new versions of a module as more features are added to it. Not all particle systems used in production will need to be updated to the latest version of the module because there may be breaking changes where particle systems using the old version of the module stop working when the module is upgraded. Normally, you can manage versions using a version control system. However, when developing Niagara modules, we may need a particular particle system to keep using its old version of the module in production and switch to new versions when desired.

Niagara maintains its own internal versioning system for modules, where you can update version numbers as you update the modules and ensure that we can select what version of the module will be used in production.

Let's assume that the **Presence Detector** module we worked on in the earlier chapters needs to be updated, and we want to create a new version with, say, green colored particles instead of the red ones we have now. However, we do not want to automatically update the existing particle system used in the levels once we publish the new module with green colored particles because some of the particle systems may have been approved and are being used in production.

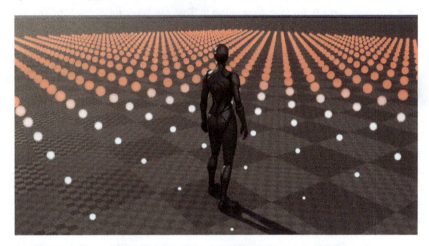

Figure 8.43: The Presence Detector Module created earlier

Just to ensure that the particles you create look like what is shown in *Figure 8.42*, use the following values for **Spawn Particles in Grid**:

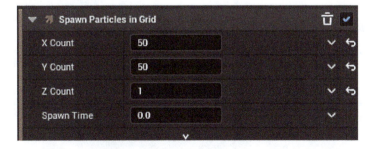

Figure 8.44: Adjusting the X Count, Y Count, and Z Count values

And use the following values for **SpriteSizeMultiplier** in **Presence Detector**:

Figure 8.45: Setting SpriteSizeMultiplier to 0.4

We will now start the process of making changes to the **Presence Detector** module and upgrading its version. We will manage this version upgrade with the versioning system built in Unreal. We need to enable the versioning system before we start using it. This can be done by clicking the **Versioning** button at the top of the Niagara Module script editor.

Figure 8.46: Click the Versioning button to enable the versioning system

A new **Versioning** panel will appear with a message that will warn you about the pros and cons of enabling the versioning system. It warns you that while you can make changes in the module without breaking existing usages, it will also mean that users will have to automatically update to new versions. Click on the **Enable versioning** button to dismiss this warning.

Figure 8.47: Click on the Enable versioning button to dismiss this dialog

The Niagara Module Script will now have some additional text next to it indicating the version number of the module.

Figure 8.48: The Versioning panel lets you add versions and manage them

For the first version, it will say something like **Niagara Module Script - Version 1.0**

If you have only 1 version, it is an exposed version. This means that if you make changes to the version, they will automatically be pushed to existing usages of the module across your system.

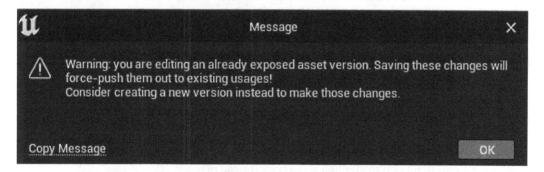

Figure 8.49: The Module editor will label the module script as Version 1.0

Niagara will warn you if you are making changes to exposed versions.

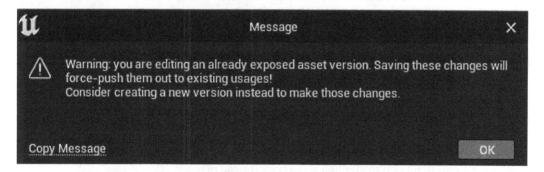

Figure 8.50: Niagara warns you if you start editing an exposed version

To prevent your changes from breaking existing usages, create a new version by clicking on the +**Add version** button in the **Versioning** panel. You can make changes in that version without worrying about breaking anything. Make sure to add the **Change Description** information for this new version, as shown in *Figure 8.47*.

Figure 8.51: Adding version 1.1

To switch to a new version for editing, just select it in the **Versioning** panel hamburger menu. All versions will be listed, and you can choose the version that you want to be active for editing.

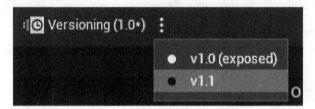

Figure 8.52: Switching the version through the hamburger menu

In this case, we have chosen **v1.1**, which is not exposed, so any changes you make here will not be automatically pushed to existing usages. The editor label will change to the working version, as shown in *Figure 8.52*.

Figure 8.53: The Module editor now displays the new version number

Let's make a few changes to our **Presence Detector** module. We will change the linking order of the **Make Linear Color** input pins to give us a green color instead of red.

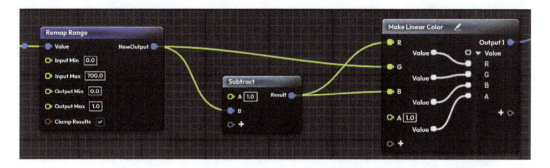

Figure 8.54: Rearranging the input pins of the Make Linear Color node

> **Note**
>
> Compare *Figure 8.53* with *Figure 7.56* of *Chapter 7*, for the original flow of the pins.

While you can save these changes to the module, these changes will not be automatically pushed to the **NS_PresenceGrid** particle system. We have to change the active version used in our particle system manually.

To make *v1.1* our active version in our particle system, select the **Presence Detector** module in the **Emitter Overview** node. In the **Selection** panel, you should see an intertwined arrow icon next to the label.

Clicking on the intertwined arrows icon will open a menu showing the available versions. The active version will be indicated by a * next to it. You can now switch to *v1.1*.

Figure 8.55: Switching to the newer version of the Presence Detector
module in the NS_PresenceGrid particle system

The particle system's behavior will change to show the effect of the new version of the custom module. You can switch back to the old version by again clicking on the intertwined arrows icon and choosing *v1.0*.

Figure 8.56: The NS_PresenceGrid particle system is now updated

The publishing and versioning workflow makes it easy to manage a parallel workflow where the level design and the particle system asset creation can happen at the same time.

With this, we come to the end of this chapter on **Local Modules** and versioning.

Summary

Using **Local Modules** is a convenient and quick way to get custom modules working, and they are very good for prototyping. Custom modules need to be managed as we continue to develop and extend them. By understanding Niagara's versioning system, we can ensure that the custom modules and systems remain compatible with future updates to the framework. This can help minimize the risk of errors and unexpected behavior when upgrading to new versions of the modules.

In this chapter, we saw a different type of custom module called **Local Module**. We saw how to use the versioning system in Niagara to keep track of new versions of modules. In the next chapter, we will learn about Events and Event Handlers, which let different emitters in a particle system pass information between them.

9

Events and Event Handlers

In this chapter, we will learn about Events and Event Handlers. This feature in Niagara allows different emitters to communicate with each other to drive effects dependent on each other. This makes it possible to build interesting effects. We will see some of the particle behaviors we can build with Events and Event Handlers. We will understand this through tutorials in which we will create fireworks, raindrops, and rocket trail behaviors.

In this chapter, we will cover the following topics:

- What are Events and Event Handlers?
- Tutorial – how to create fireworks

Technical requirements

Unreal Engine 5.1 and above is required for this chapter. The installation procedure was already explained in *Chapter 1*.

You can find the project we worked on in this book here:

`https://github.com/PacktPublishing/Build-Stunning-Real-time-VFX-with-Unreal-Engine-5`

What are Events and Event Handlers?

As we write complex particle systems, for them to work, we will need to have multiple emitters in a system interact with each other. For example, when creating a firework effect, when the particles that streak into the sky from the ground die, they should spawn a secondary sparkling effect. With raindrops, when each water particle hits the ground, it should spawn a splashing particle effect. Events and Event Handlers enable us to achieve these kinds of effects. Particles generate specific Events that occur in the lifetime of the particle. Event Handlers listen for those Events and respond to them. The response may be in the form of property or behavior changes or the spawning of particles. For example, in the case of raindrop effects, the water particle, on hitting the ground, generates a **Collision** event. This triggers the Event Handler to spawn secondary particles, which create a splashing effect.

Events work only with CPU simulations. We also need to enable **Require Persistent IDs** in our **Emitter Properties** to enable the indexing of particles. Without persistent IDs, we would not be able to keep track of particles.

There are three event modules available in Niagara:

- **Location**
- **Death**
- **Collision**

These three modules, when added to the **Emitter** node, are used to generate the corresponding event. The **Location** module, for example, when added, will let you use **Generate Location Event**. Without the module, the Events cannot be generated.

Let us now understand how an event is received. To receive an event, you need to add an **Event Handler** stage to your emitter. Once the **Event Handler** stage is added to your **Emitter** node, you can add a module to receive the event that matches the generated event. For example, if you generated a **Location** event, you would need to have a **Receive Location Event**.

As you would expect, this requires two emitters in a system: one to generate the event and the other to handle the event. To better understand this, let us create a Niagara System that incorporates Events and Event Handlers.

Tutorial – how to create fireworks

In the following tutorial, we will create fireworks using all three event modules. Let's get started:

1. Let us create a new folder called `EventDispatcher` and in this folder, let us create two emitters. Let us name the first emitter `Receiver` and the second emitter `Sender`.

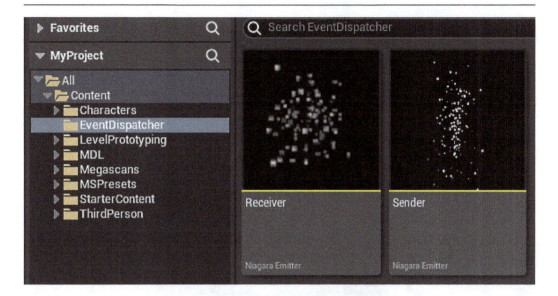

Figure 9.1: Creating two emitters called Receiver and Sender

We will use the **Fountain** template for the **Sender** emitter.

Figure 9.2: The Fountain template for Sender

And we'll use the **Omnidirection Burst** template for the **Receiver** emitter.

Figure 9.3: The Omnidirectional Burst template for Receiver

We have named these emitters in this way so that we can understand which emitter will generate the Events (**Sender**) and which emitter will receive them (**Receiver**). Feel free to use any names you find convenient as they have no bearing on the functionality.

Figure 9.4: Creating the NS_EventHandlerExample Niagara System
incorporating the Reciever and the Sender emitters

2.   We will now create a Niagara System that incorporates these emitters. We will name this system NS_EventHandlerExample. We will add our **Sender** and **Receiver** emitters to this System by selecting them from the **Parent Emitters** tab when we create the system.

Figure 9.5: The Niagara NS_EventHandlerExample particle system
icon along with the Reciever and Sender emitter icons

3. As we have already seen in the previous examples of creating a system, once we open the **NS_EventHandlerExample** particle system in the Niagara editor, we will have the **Emitter** nodes and the **System** node with the selected template-provided modules, as seen in *Figure 9.6*.

We will enable Event and Event Handler functionality in this particle system. In Unreal Engine 4, the Event Handlers were available and visible in the **Emitter** nodes. In Unreal Engine 5, we do not see them by default. We must complete an extra step to activate the Events and Event Handlers, which we will do later in this chapter.

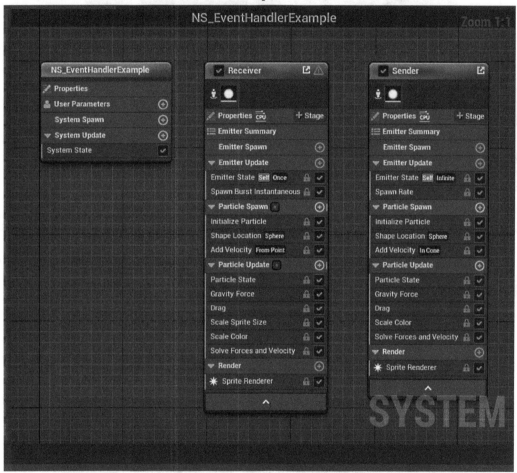

Figure 9.6: Editing the NS_EventHandlerExample particle system

4. For creative purposes, we will make a small modification to our **Receiver** emitter. In the **Timeline** panel, the **Receiver** emitter will have a keyframe at **Frame 0**. We will select and delete that keyframe.

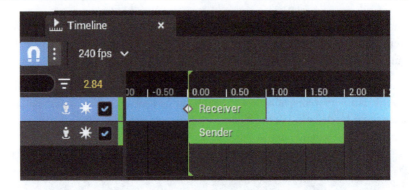

Figure 9.7: Deleting the keyframe at Frame 0 from the Receiver emitter

This will enable the emitter to have continuously spawning particles instead of an instantaneous burst. This is a creative call to make our fireworks look right.

Death Event

In **Sender**, let us generate our first event:

1. On the **Particle Update** label, click on the + sign and choose **Generate Death Event**. The particle system compilation may fail here, but don't panic as we are going to fix it in a few steps.

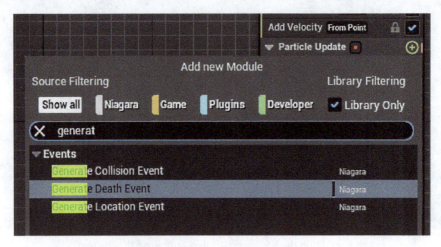

Figure 9.8: Adding the Generate Death Event

2. This will add the **Generate Death Event** module to the **Sender** emitter node.

Figure 9.9: The Generate Death Event module added to the Sender node

3.   Select the **Generate Death Event** module to see the relevant details in the **Selection** panel.

Figure 9.10: Details of the Generate Death Event module in the Selection Panel

We will have to make a few changes in the **Sender** node to address any compilation failures and for Events to work.

We'll do it with the following steps:

1.   Let us choose the **Properties** module in the **Sender** node.

Figure 9.11: Modifying the Properties module in Sender

2. In the **Selection** panel for this module, ensure **Sim Target** is set to **CPUSim** and check the **Requires Persistent IDs** option. Neither Events nor event dispatchers will work if these conditions are not met.

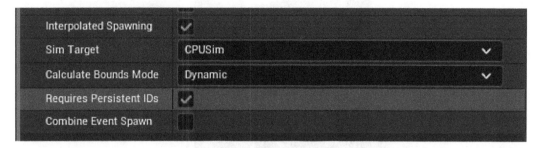

Figure 9.12: Sim Target set to CPUSim and Requires Persistent IDs checked

3. Let us now tweak the properties of the **Sender** emitter. This is a creative change and is optional. We are doing this so that we can see the effects on the **Emitter** node more clearly.

    In the **Initialize Particle** module, we will set the particle's **Lifetime Mode** setting to **Random** with **Lifetime Min** set to 0.25 and **Lifetime Max** set to 0.75.

Figure 9.13: Tweaking Lifetime values for better visibility of the particles

This will shorten the fountain as shown in *Figure 9.14*.

Figure 9.14: The fountain with the tweaked values is now much shorter

4.   We will also reduce the **SpawnRate** option in the sender to 5 . 0, again, as a creative choice. We do not want too many fireworks launched.

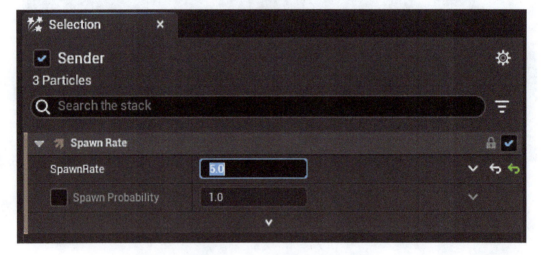

Figure 9.15: Reduce SpawnRate to 5.0

Now, let us move our attention to the **Receiver** node:

1.   We will add the **Event Handler** stage to the **Receiver** node by clicking on the + **Stage** button on the **Properties** module. This adds an **Event Handler** section to the emitter and an **Event Handler Properties** node to it.

Figure 9.16: Adding the Event Handler stage to the Receiver node

We need to configure the **Event Handler Properties** parameters to enable the receiver to **Receive** events.

Figure 9.17: The Event Handler Properties module showing up in the Receiver node

2. In **Event Handler Properties**, let us change the source to **Death Event**. We will see the **Death Event** option only if we have added **Generate Death Event** to the **Sender** node.

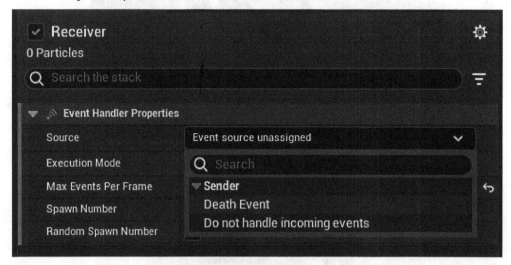

Figure 9.18: Changing Source to Death Event

3. Let us also set **Execution Mode** to **Spawned Particles**. This will make the event script run only on the particles that were spawned in response to the current event in the emitter.

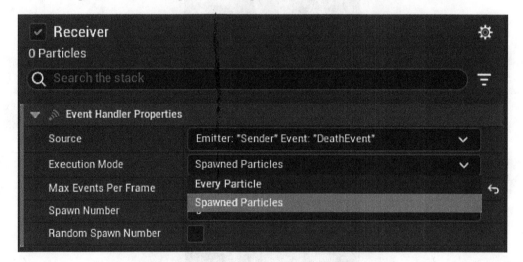

Figure 9.19: Changing Execution Mode to Spawned Particles

4. Now, in the **Spawn Number** field, set the number of particles spawned to 5. These are the number of particles that will be spawned by the **Receiver** emitter when the **Sender** emitter's particle dies.

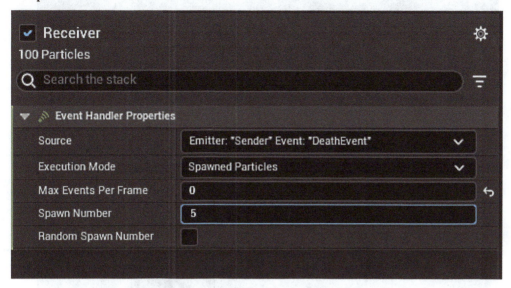

Figure 9.20: Setting Spawn Number

5. Let us now add **Receive Death Event** to the **Receiver** emitter. This is the module that will receive the death event from the **Sender** emitter's **Generate Death Event** module.

Figure 9.21: Adding Receive Death Event to the Receiver emitter

6. We also need to change the **Life Cycle Mode** setting of **Receiver** from **Self** to **System**. This will allow the life cycle of the **Receiver** emitter to be controlled by the system rather than the emitter itself.

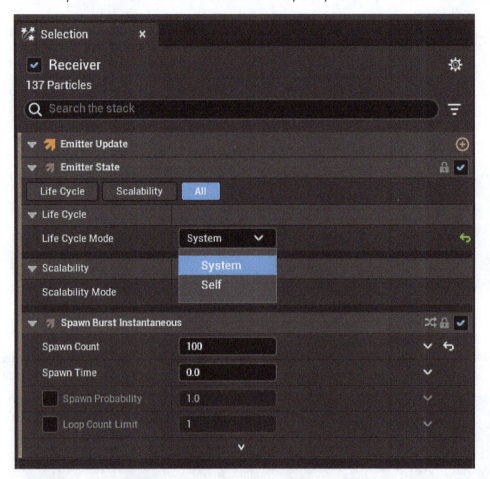

Figure 9.22: Setting the Life Cycle Mode option to System

7. To make it easy to differentiate between the **Sender** particles and the **Receiver** particles, let us change the color of the **Receiver** sprites to red.

Figure 9.23: Changing the color of Receiver sprites to red

You should now be able to see the Events and Event Handlers in action in the **Preview** window.

Sender's white particles will emit like a fountain into the air, and after they die, **Event Handler** will spawn red **Receiver** particles.

Figure 9.24: The particle system with red-colored Receiver sprites being
emitted on the death of the white-colored Sender sprites

Feel free to tweak and play around with this system to create a variety of effects. This effect looks like a fireworks display.

The Collision Event

Let us now check the **Collision** event and its corresponding Event handling:

1. Similar to the **Death** event, let us add the **Generate Collision Event** module to the **Sender** node. You can disable or remove the **Death** event module that we added earlier.

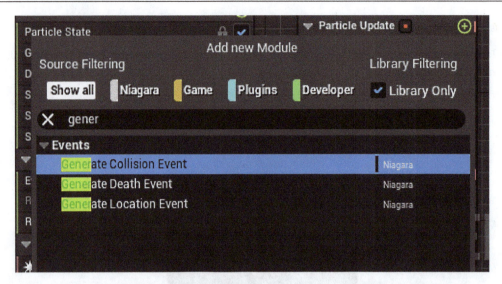

Figure 9.25: Adding a Generate Collision Event module to the Sender node

2. Niagara informs us that there is an unmet dependency, which, in this case, is the **Collision** module. We can click on the **Fix issue** button to resolve this problem.

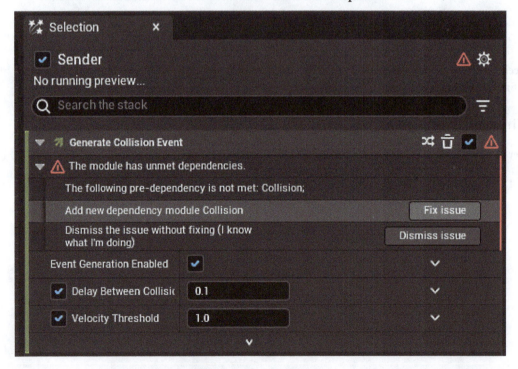

Figure 9.26: Fixing the dependency issue by adding a Collision module to Sender

3. In the **Receiver** node, add the **Receive Collision Event** module and disable the **Death Event** module.

Figure 9.27: Adding Receive Collision Event to the Receiver node

We will need to make some changes to the **Sender** node for the effect to be seen clearly.

4. We want the particles to live long enough to hit the floor and collide. To achieve this, change **Lifetime Min** and **Lifetime Max** to 1.25 and 2.5 respectively.

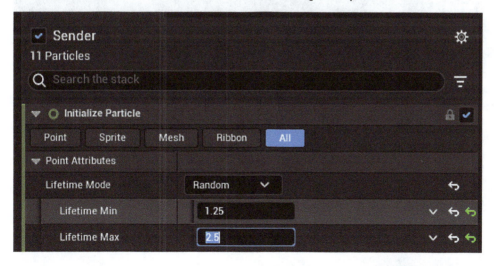

Figure 9.28: Making creative changes to the Lifetime values to make the particles live longer

5.   In the **Receiver** node's **Event Handler Properties**, change the **Source** dropdown to **Collision Event**.

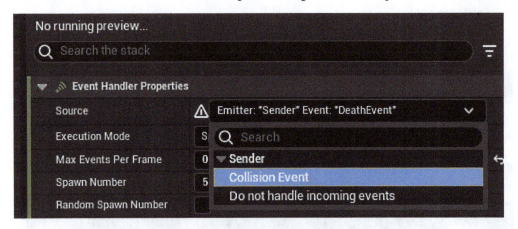

Figure 9.29: Changing Source to Collision Event in the Receiver node. The existing entry in Source is from the now-disabled Death Event module

Figure 9.30 is what you'll now see.

Figure 9.30: The Source property after being set to Collision Event

6. Note that the earlier entry of **"DeathEvent"** in the **Source** property is due to the now-disabled **Death Event** module. It will not show up in the menu. There is a warning triangle shown because we disabled the Death Event. This triangle will disappear after we select **Collision** Event.

You should see a change in the behavior of the particle system. Sender's white particles will now emit and fall to the floor. On colliding with the floor, a **Collision** event is generated causing the **Receiver** node to emit red particles.

Figure 9.31: The final raindrop effect

This effect is great to create a raindrop splash effect or welding sparks effect.

The Location Event

Let us now explore the **Location** Event and the corresponding Event Handler. It is very similar to the two Events that we just examined:

1. Let us start by adding a **Generate Location Event** module and disabling or deleting the **Collision** event in the **Sender** node.

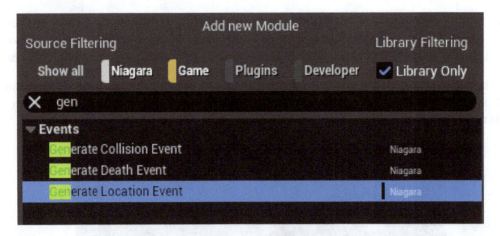

Figure 9.32: Adding a Generate Location Event module to the Sender node

2. We will also add **Receive Location Event** to the **Receiver** node. Make sure to disable or delete any other **Receive** events.

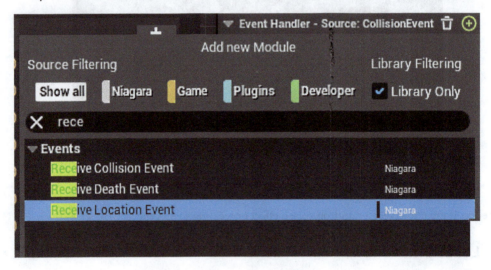

Figure 9.33: Adding Receive Location Event to the Receiver node

3.   In the **Receiver** node, change the **Source** dropdown to **Location Event**.

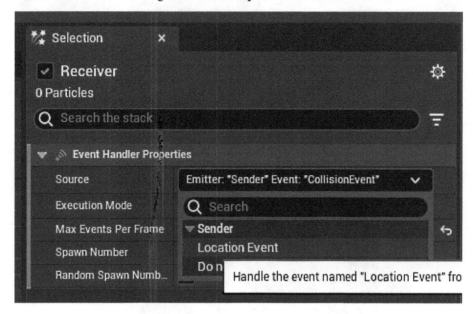

Figure 9.34: Changing the Source setting to Location Event in the Receiver node. The existing
entry in the Source dropdown is from the now-disabled **Collision Event** module

Figure 9.35 is what you'll now see.

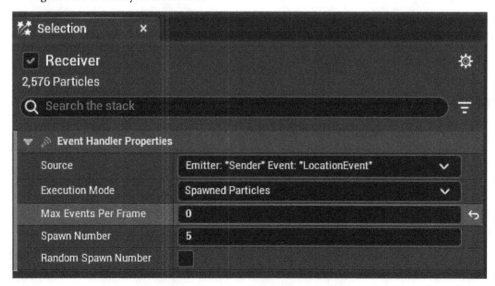

Figure 9.35: The Source property after being set to Location Event

4. Increase the number of spawned particles in the **Receiver** node as needed.

Figure 9.36: The particles now behave like a missile trail

You should see red **Receiver** particles emit and form a trail behind the white **Sender** particles. This effect is good for missile/rocket trails and similar effects.

With this, we've come to the end of the chapter.

Summary

In this chapter, we became familiar with useful features in Niagara called Events and Event Handlers. This enabled us to connect emitters and have one emitter drive the other emitter's effects. We saw some functional examples of how to implement this feature.

We have been working on the creation part of Niagara until now. During development, we also need to figure out any bugs that creep into our particle systems, as well as keep an eye on the efficiency of the particle system.

In the next chapter, we will learn some debugging tricks to help enhance the performance of a Niagara System and ways to monitor it.

10

Debugging Workflow in Niagara

Niagara particle systems are essentially abstracted pieces of code controlling the behavior and rendering of particle systems. Code has a nasty habit of having bugs in it and we end up spending a lot of time trying to debug that code. In this chapter, we will explore the process of debugging in Niagara.

We will familiarize ourselves with the **Debugger** panel, which has tools to sift through the particle system data. Then, we will look at data visualization tools that overlay debug data on our systems. We will check some performance profiling features and finally, we will learn a few important console commands to add to our debugging toolset.

This is what we'll be covering in the chapter:

- Exploring the Niagara Debugger panel
- Debug Drawing
- Performance profiling
- Debug console commands

Technical requirements

You can find the project we worked on in this book here:

`https://github.com/PacktPublishing/Build-Stunning-Real-time-VFX-with-Unreal-Engine-5`

Exploring the Niagara Debugger panel

Niagara provides you with a debugger to review particle systems running in your level. The debugger has a bunch of tools to help you look at the simulation data in detail to pinpoint any issues in your particle system.

You can turn on the **Debugger** panel by going to **Tools** > **Debug** > **Niagara Debugger**.

This will open the dockable **Niagara Debugger** panel. As with any panel in Unreal, you can dock it to any convenient position.

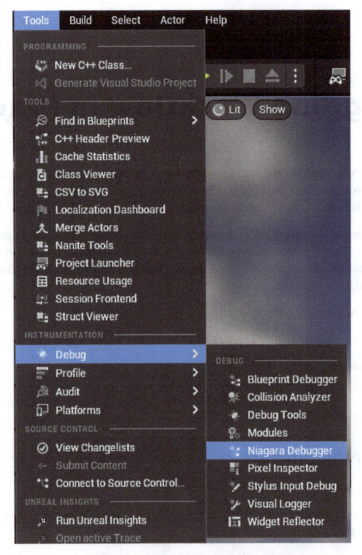

Figure 10.1: Opening Niagara Debugger

The **Niagara Debugger** panel has three major areas:

- The **Particle System Playback Options** toolbar
- Debugger tabs
- Debugger options

This is depicted in the following figure (*Figure 10.2*):

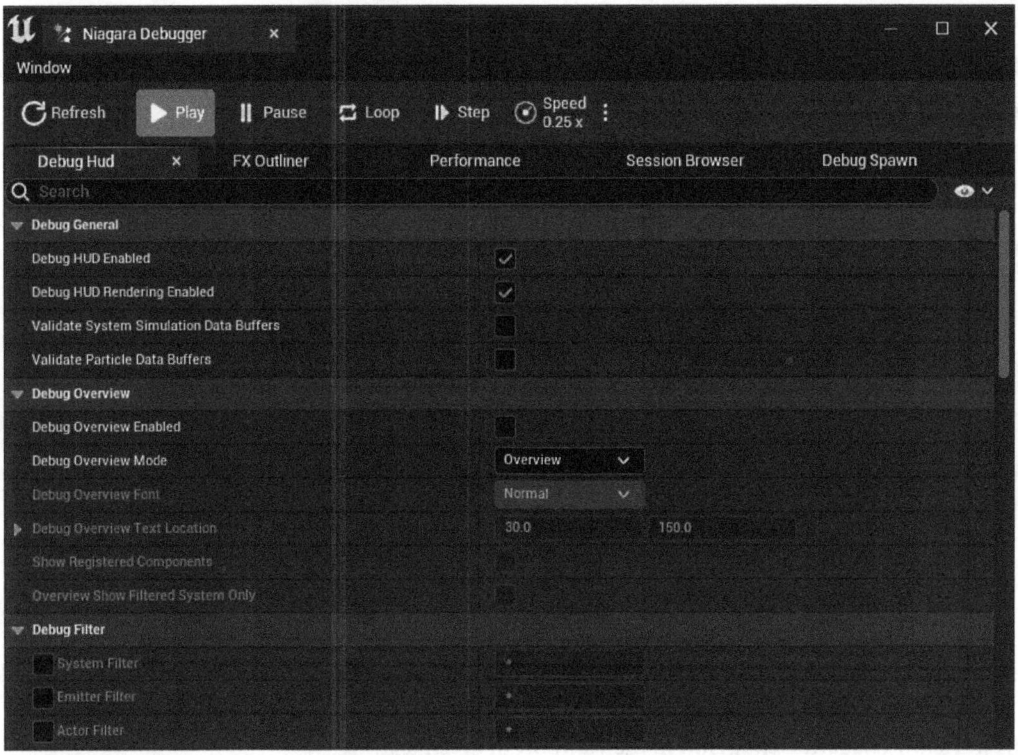

Figure 10.2: The Niagara Debugger panel

Each of the tabs contains further options. Please be aware that this layout may change in upcoming versions of Niagara.

The Particle System Playback Options toolbar

For diagnostic purposes, we may sometimes need to pause the particle system, play it at a slower rate, or step through each frame. The **Particle System Playback Options** toolbar (*Figure 10.2*) will allow us to control the playback of the Niagara Systems as required.

The toolbar has the following buttons:

- **Refresh**: This lets you refresh the particle system setting to fix out-of-sync behavior
- **Play**: This lets you play Niagara simulations in the active level
- **Pause**: This lets you pause Niagara simulations in the active level
- **Loop**: This can be used to make single-firing particle systems loop

- **Step**: This lets you step forward through all particle systems by one frame

- **Speed**: This lets you change the playback speed of the particle systems

Now that we've talked about the playback buttons, let us have a detailed look at the first tab panel, **Debug Hud**.

Debug Hud

To enable Niagara Debugger to start showing debugging info, check the **Debug HUD Enabled** option in the **Debug General** section if it is not already enabled:

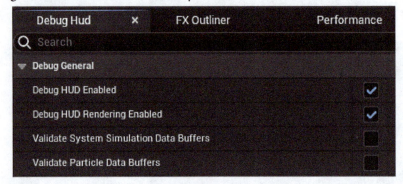

Figure 10.3: The Debug HUD tab panel

This will show **Niagara DebugHud** in the viewport in green text:

Figure 10.4: Niagara Debug Hud currently showing no data

If you don't see any green text, check the **System Filter** checkbox in the **Debug Filter** section:

Figure 10.5: Enable the System Filter in the Debug Filter section

We have to turn on a few more properties for us to see detailed particle system data.

In the **Debug Overview** section, turn on **Debug Overview Enabled**. Set **Debug Overview Mode** to **Overview**:

Figure 10.6: The Debug Overview section

This will show us the following properties:

- **TotalSystems**
- **TotalScalabilty**
- **TotalEmitters**
- **TotalParticles**
- **TotalMemory**

In *Figure 10.7*, we can see that the total number of Niagara Systems in our level is **1** and there is **1** emitter with **125000** particles simulated in total, which is using **50.42** MB of memory. We have not applied any scalability setting to our particle system and hence **TotalScalability** reads **0**. Scalability settings are added in **Project Settings** > **Plugins** > **Niagara**.

Figure 10.7: Niagara DebugHud after turning on Debug Overview
Enabled and setting Debug Overview Mode to Overview

You can modify the font and the **Debug Overview** text location by changing the appropriate options. Here, we will leave them at the default settings. Turn on **System Filter** in the **Debug Filter** section.

If you are seeing a lot of data in **DebugHud** and want to filter only selected data, you can replace the * symbol in the **System Filter** input box with the string you want to filter out. For example, type in `*sparks*` to display objects having the `sparks` substring in them.

Figure 10.8: Replace the * in the System Filter input box with the string that needs to be filtered

We can also filter emitters using **Emitter Filter** if we have multiple emitters in our system. **Emitter Filter** is located just below **System Filter**, as seen in *Figure 10.8*.

We will be debugging the `NS_PresenceGrid` particle system that we created in *Chapter 7*.

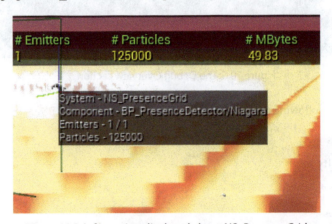

Figure 10.9: Information displayed about NS_PresenceGrid

Because we turned on the System filter with the * wild card, the display will now change to show detailed information about all particle systems in your level. *Figure 10.9* shows the information that will appear on the screen for `NS_PresenceGrid`.

We can modify the details shown on the screen by changing various properties in the **Debug System** section, as shown in *Figure 10.10*. The **Debug System** section displays detailed information about each particle system present in the level next to it in the viewport.

Figure 10.10: The Debug System section

Let's turn on **System Show Bounds**. This will show the bounds of the particle system as a box with red edges. If you are using GPU particles, you can use the bounding box information to troubleshoot the particle system disappearing due to camera frustum occlusion turning off the particle system.

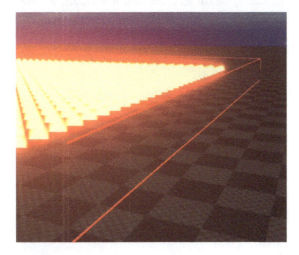

Figure 10.11: The red colored bounding box, which shows up after turning on System Show Bounds

The appearance of the debugger text on the system can be changed in **System Text Options** by changing the **Font** size, **Horizontal Alignment**, and **Vertical Alignment**.

The **Debug Particles** section gives detailed information on individual particles. Turning on **Show Particle Attributes** will turn on the display of the individual particle information. If you need the debugger to show additional particle attributes in addition to the default **Index**, such as **Position** and **Color**, you can add them in the **Particle Attributes** array, which is present in the **Debug Particles** section. In Unreal version 5.1, you may have to manually add **Position** and **Color** attributes as they may be missing. Make sure to copy the exact attribute name. You can get the names of the attributes available in the **Parameters** panel and look for parameters in the **PARTICLES** namespace. In this case, we have added the **Age** attribute:

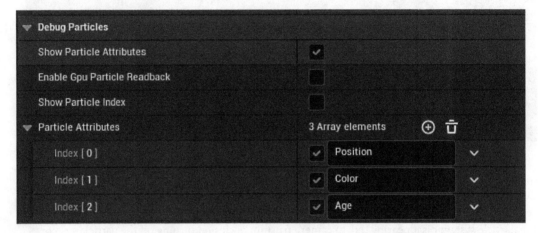

Figure 10.12: Adding the names of the Particle Attributes to the Particle Attributes array. These attributes will show up in the debug display

You will see the debug display change to show the **Age** data in addition to **Position** and **Color**:

System - NS_PresenceGrid
Component - BP_PresenceDetector/Niagara
Emitters - 1 / 1
Particles - 125000
Emitter (Empty)
Particle(0) Age(4.10) Color(1.00, 0.00, 0.00, 1.00) Position(900.00, -800.00, -6.00)
Particle(1) Age(4.10) Color(1.00, 0.00, 0.00, 1.00) Position(1000.00, -800.00, -6.00)
Particle(2) Age(4.10) Color(1.00, 0.00, 0.00, 1.00) Position(1100.00, -800.00, -6.00)
Particle(3) Age(4.10) Color(1.00, 0.00, 0.00, 1.00) Position(1200.00, -800.00, -6.00)
Particle(4) Age(4.10) Color(1.00, 0.00, 0.00, 1.00) Position(1300.00, -800.00, -6.00)
Particle(5) Age(4.10) Color(1.00, 0.00, 0.00, 1.00) Position(1400.00, -800.00, -6.00)
Particle(6) Age(4.10) Color(1.00, 0.00, 0.00, 1.00) Position(1500.00, -800.00, -6.00)
Particle(7) Age(4.10) Color(1.00, 0.00, 0.00, 1.00) Position(1600.00, -800.00, -6.00)
Particle(8) Age(4.10) Color(1.00, 0.00, 0.00, 1.00) Position(1700.00, -800.00, -6.00)
Particle(9) Age(4.10) Color(1.00, 0.00, 0.00, 1.00) Position(1800.00, -800.00, -6.00)
Particle(10) Age(4.10) Color(1.00, 0.00, 0.00, 1.00) Position(1900.00, -800.00, -6.00)
Particle(11) Age(4.10) Color(1.00, 0.00, 0.00, 1.00) Position(2000.00, -800.00, -6.00)
Particle(12) Age(4.10) Color(1.00, 0.00, 0.00, 1.00) Position(2100.00, -800.00, -6.00)
Particle(13) Age(4.10) Color(1.00, 0.00, 0.00, 1.00) Position(2200.00, -800.00, -6.00)
Particle(14) Age(4.10) Color(1.00, 0.00, 0.00, 1.00) Position(2300.00, -800.00, -6.00)
Particle(15) Age(4.10) Color(1.00, 0.00, 0.00, 1.00) Position(2400.00, -800.00, -6.00)

Figure 10.13: The debug display showing Position, Color, and Age attributes for each particle

If you do not see the particle attributes, check whether **System Debug Verbosity** is **Basic** or **Verbose**.

FX Outliner

The second tab, named **FX Outliner**, lets you capture Niagara simulation data for analysis.

You can set the number of frames to capture in the **Delay** input box. Here, we will keep the default value of **60**. Check that the **Perf** button is highlighted. This will ensure that the **Performance** data is captured. Click on the **Capture** button to start capturing the data:

Figure 10.14: The FX Outliner tab panel

We will not modify any other settings in **FX Outliner**. Since the default for **View Mode** was **State**, we see the captured data as shown in *Figure 10.15*. The data is shown with the following hierarchy:

- **World**: This is the level name. In our case, it is **ThirdPersonMap**. Here, the inline data tells us that the data was captured from **Editor** as the source. Other sources may be **Game** or **PIE**. Next, it tells us whether the game was running as a standalone, dedicated server, or a client. This is useful while debugging multiplayer games. In our case, we were running as **Standalone**. The next option tells us whether **Begin Play** has been activated. Since we were debugging it in **Editor**, and not running the game, it returns **False**. The last number tells us the number of systems captured. This number may be affected by any filters that you set.

- **System**: This is the Niagara System for which performance was captured. The inline data informs us that **1** system was found. This number is affected by the filters (if any) set.

- **System Instance**: This is the particular instance of the Niagara System present in our level. The inline data in this case informs us that the **Pooling** method used here was **None**. Pooling methods are used when we want to reuse objects rather than allocate new ones. If a pooling method was used in the system, it will display **InUse** or **FreeInPool**. Pooling methods are generally set when you spawn a Niagara System through blueprints. Next, it tells us the **Execution** state. In this case, it is **Active**. Other states are **Inactive** and **Complete**. Finally, it shows the number of emitters matching the current filter.

- **Emitter**: This is the emitter present in the system. The Inline data informs us whether the emitter is **Active, Inactive**, or **Complete**. In our case, it is **Active**. It also shows whether it is a **CPU** or **GPU** emitter. Finally, it shows the number of particles that were emitted by the emitter that are alive.

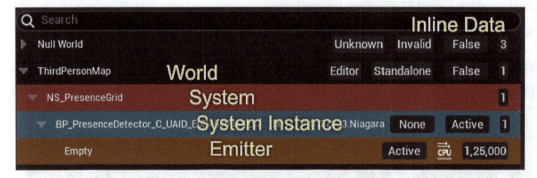

Figure 10.15: FX Outliner data in the State View Mode

In **Performance View Mode**, different Inline Data related to performance is shown. In total, there are eight blocks of data shown. Each block has two values. The left value is the game thread cost, and the right value is the render thread cost of the system. The game thread is the one that calculates the behavior part of the particle system, while the render thread is the one responsible for displaying the particles on screen. The values are in microseconds:

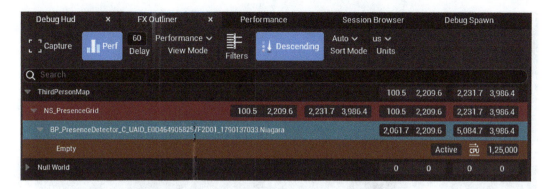

Figure 10.16: FX Outliner data in Performance View Mode

Let's look at the values captured over the duration of the Delay.

This is what **World (ThirdPersonMap)** captured (from left to right):

- Average total frame time for all effects in the world
- Maximum total frame time for all effects in the world

This is what **System (NS_PresenceGrid)** captured (from left to right):

- Average per-instance cost for this system
- Maximum per-instance cost for this system
- Average total cost for all instances of this system
- Maximum total cost of all instances of this system

This is what **System Instance** (the **BP_PresenceDetector** object) captured (from left to right):

- Average cost for this instance
- Maximum cost for this instance

This can help you pinpoint at which point in the hierarchy the efficiency issues may be present. If the render thread cost is high, you might want to check your shaders.

The **Performance** tab is very much in development, so expect it to change drastically. The main button here is the **Run Performance Test** button, which will run the test for the number of frames indicated on the button. Click on the button and check the **Output Log** panel.

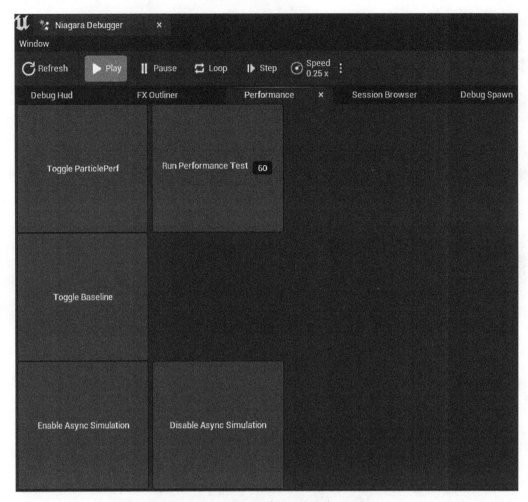

Figure 10.17: The Performance tab panel

In **Output Log**, you will see a bunch of data printed out. This data may seem meaningless, but it is in **Comma Separated Values (CSV)** format. Copy this data and save it as a .CSV file:

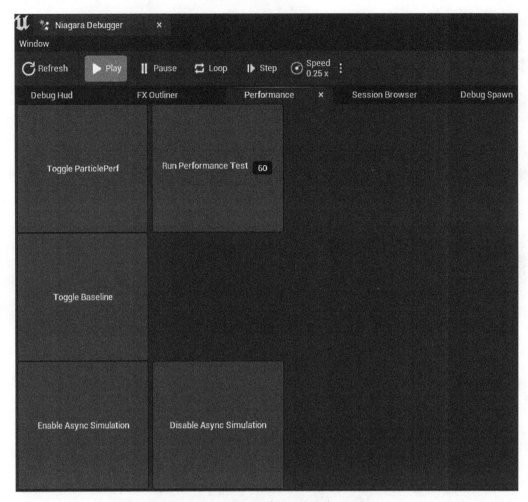

Figure 10.18: Pressing the Run Performance Test button in the Performance
tab panel prints out CSV formatted data in the Output Log

Open the CSV file in any spreadsheet of your choice. The data will show up as seen in *Figure 10.19* and can then be further evaluated:

**** Particle Performance Stats								
Name	Average PerFrame GameThread	Average PerInstance GameThread	Average PerFrame RenderThread	Average PerInstance RenderThread	NumFrames	Total Instances	Total Tick GameThread	Total Tick Con
** Per System Stats								
NS_PresenceGrid	0	0	3871	1935	0	0	0	
**** Particle Performance Stats								
Name	Average PerFrame GameThread	Average PerInstance GameThread	Average PerFrame RenderThread	Average PerInstance RenderThread	NumFrames	Total Instances	Total Tick GameThread	Total Tick Con
** Per System Stats								
NS_PresenceGrid	0	0	3871	1935	0	0	0	
**** Particle Performance Stats								
Name	Average PerFrame GameThread	Average PerInstance GameThread	Average PerFrame RenderThread	Average PerInstance RenderThread	NumFrames	Total Instances	Total Tick GameThread	Total Tick Con
** Per System Stats								
NS_PresenceGrid	0	0	3871	1935	0	0	0	

Figure 10.19: Saving the data from the Output Log as a .csv file and opening in a spreadsheet

Session Browser is used to debug instances of the level along with the Niagara Systems running on another system. For example, if you are developing a console game, you can run the level on the console, and **Session Browser** will display the console device if it is connected to the system on which the debugging is being done. You can then capture the debugging information from the console device by selecting it in **Session Browser**. In our case, we can see the single session running on our local device:

Figure 10.20: Session Browser

Debug Spawn is a new feature introduced in Unreal Engine 5 to Niagara Debugger. Using this feature, we can spawn and kill particle systems in a level without having to build elaborate spawning mechanisms. This lets you figure out the overhead generated by just the particle systems.

Figure 10.21: The Debug Spawn tab panel

To spawn particles using this feature, add the particle systems you want to spawn in the **Systems to Spawn** array. You can specify the location where you would want the particle system to spawn. With **Debug HUD** on, click on the **Spawn Systems** button. This will spawn the particle system in the level, which can help debug the particle system independent of any dependencies.

To kill the spawned systems, click on the **Kill Existing** button.

Figure 10.22: Adding the NS_PresenceGrid particle system to the Systems to Spawn array

These are some of the options you have to debug instances of particle systems in your level. In the next section, we'll look at some available tools to debug the system while you are creating it in the Niagara Editor.

Debug Drawing

For some modules, it is difficult to visualize the properties that the module applies to the particle behavior. It may also not be very clear, from observing the resultant particle behavior, as to what is the contribution of a particular module to that behavior.

In these cases, having the capability of representing such properties in a way that is easy to debug is very helpful. For example, the direction of movement of a particle, the direction in which the force is being applied, and so on can be drawn on top of the particle system render as debug lines to make it clear to the user what exactly is happening in the system. The Debug Drawing feature is available on a few select modules.

We'll take a look at a few examples.

For our first example, let us create an emitter called NE_DebugDemo using the fountain template.

Figure 10.23: Creating the NE_DebugDemo emitter using the fountain template

We will add a **Collision** module to this emitter in the **Particle Update** group to enable floor collisions for this emitter.

Figure 10.24: Adding the Collision module to the emitter node

The **Collision** module will show up under the **Particle Group** in the emitter node, as shown in *Figure 10.25*:

Figure 10.25: The Collision module added to the NE_DebugDemo emitter overview node

The particles should now be bouncing off the floors. You will notice a small eye icon on the **Collision** module. This enables the Debug Drawing for this module. Click on the eye icon, which will turn it green, and turn on the CPU raytracer debug draw, which is used to calculate the collision of the particles with the world.

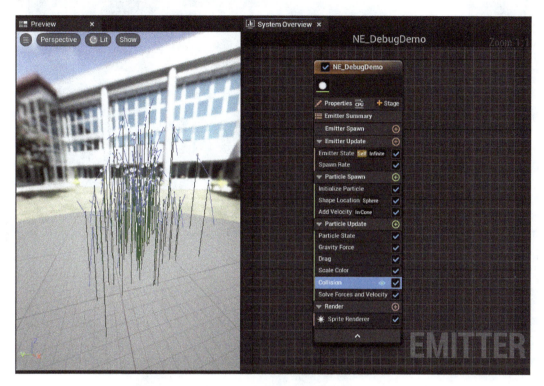

Figure 10.26: The debug draw turned on in the Collision module by clicking on the eye icon

Now for our second example, let us disable the **Collision** module and add a **Vortex Force** module in the **Particle Update** group. Press the eye icon on the **Vortex Force** module to turn on Debug Drawing for that module. You should be able to see red colored lines indicating the direction of the vortex force. The lines are not clear as the vortex force is small.

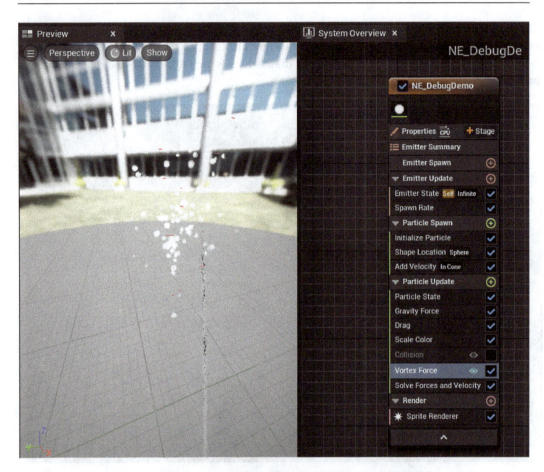

Figure 10.27: Adding the Vortex Force module and turning on debug draw. Small
red lines can be seen. The Collision module has been disabled here

Let us change **Vortex Force Amount** to 2000 so that we can see the debug lines clearly:

Figure 10.28: Setting Vortex Force Amount to 2000

You will notice in the **Preview** window that the debug lines are now longer and can be seen clearly. The length of the debug line is directly proportional to the value of **Vortex Force Amount**, which helps in diagnosing any unexpected behavior we might be having when developing a particle system.

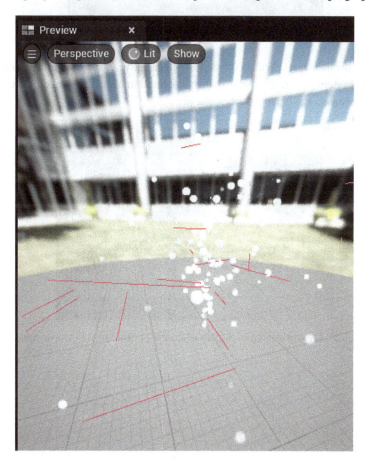

Figure 10.29: The debug lines are longer in proportion to the increased Vortex Force Amount

Debug Drawing is available only on a few select modules, where it can be useful to diagnose and troubleshoot any issues you may be having with a particle system.

Performance profiling

Another way that can help us optimize a particle system is to visualize the impact on the performance of each module in the particle system. This helps us identify modules that are taking more time to evaluate and focus on optimizing them.

In Unreal Engine 5, a new **Performance** button was added to the Niagara Editor toolbar to help us visualize the impact of each module on the performance of the particle system:

Figure 10.30: The Performance button

This button toggles the display of performance information at a granular level for each of the modules in the **Emitter** and **System** nodes. The performance information changes based on the options chosen in the dropdown menu on the **Performance** button:

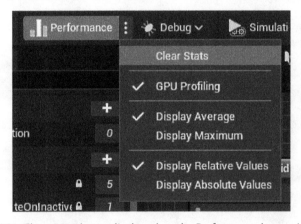

Figure 10.31: Choosing what to display when the Performance button is pressed

By default, the **Performance** button shows the average and the relative values, as shown in *Figure 10.32*:

Figure 10.32: The nodes showing average and relative values

When showing maximum and absolute values, the display is as shown in *Figure 10.33*:

Figure 10.33: The nodes showing Maximum and Absolute values

The displayed information changes as per the module for which the information is being displayed.

The main labels (such as **Emitter Update**, **Particle Spawn**, and **Particle Update**) show the total cost of the script and its modules. The module calls may not add up to 100% as the script itself has some overhead.

The individual modules show the cost of each module. When showing up as relative values, the percentage displayed is with respect to the parent script. For example, in the case of the **Grid Location** module, it is 20.9% of the **Particle Spawn** script (as seen in *Figure 10.32*).

Using the **Performance** display helps us get a quick overview of our emitters and quickly spot any inefficient modules.

Debug console commands

While the Niagara Debugger is an excellent tool, we may not need all the features it has. We may want to use the debug features at runtime without the interface, or we may want to be able to log them during gameplay testing, or we may even want to trigger the debug features at certain points in our game to debug specific areas in our games.

We can do all of this and much more with the help of console commands. These commands can be typed in the Unreal Editor console located in the **Output Log** panel:

```
fx.Niagara.Debug.GlobalLoopTime
fx.Niagara.Debug.Hud
fx.Niagara.Debug.KillSpawned
fx.Niagara.Debug.PlaybackMode
fx.Niagara.Debug.PlaybackRate
fx.Niagara.Debug.SpawnComponent
fx.Niagara.DebugDraw.Enabled
```

| 2.00 |

>_ Cmd ⌄ | fx.niagara.Debug |

Fig. 9.34: Typing the Niagara debug commands in the console

It can also be triggered through the **Execute Console Command** node in **Blueprints**:

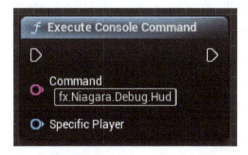

Figure 10.35: Using the Execute Console Command node in Blueprints to trigger the debug commands

Let us have a look at an example of a console command:

```
fx.Niagara.Debug.Hud Enabled=1 OverviewEnabled=1
```

This command by itself can be seen here:

```
fx.Niagara.Debug.Hud
```

Next, chained to the command, is the argument that will enable the HUD (0 to disable). This and the subsequent commands are to be added to `fx.Niagara.Debug.Hud`:

```
Enabled=1
```

The following argument enables or disables (0 to disable) the main overview display:

```
OverviewEnabled=1
```

The subsequent argument sets the system verbosity level. 0 corresponds to **None**, 1 to **Basic**, and 2 to **Verbose**:

```
SystemDebugVerbosity=0
```

This argument sets the in-world system emitter debug verbosity. 0 corresponds to **None**, 1 to **Basic**, and 2 to **Verbose**:

```
SystemEmitterVerbosity=0
```

The next argument shows all filtered systems' bounds (0 will disable it):

```
SystemShowBounds=1
```

The following argument sets the wildcard system filter for the in-world display:

```
SystemFilter=*Sparks*
```

This argument enables the `SystemVariables` visibility (0 will disable it):

```
ShowSystemVariables=1
```

The next argument would display any variable that matches `Position` and all variables that contain `Color`:

```
SystemVariables=Position,*Color
```

This is another example of a console command:

```
fx.Niagara.Debug.PlaybackRate
```

Adding `0.5` to this command, for example, will run all simulations at half speed:

```
fx.Niagara.Debug.PlaybackRate 0.5
```

You can find additional commands by typing `fx.Niagara.Debug` in the console.

Debug commands are for advanced usage, but it is a good habit to get into using the console in Unreal Engine. Many features in Unreal that are not available in the regular user interface can be accessed through console commands.

Summary

In this chapter, we learned about the debug toolset offered by Niagara to diagnose issues and performance bottlenecks in our particle systems. We also learned how to generate verbose data and save it in formats that can be further analyzed by external tools. We understood the profiling tools, which present data visually to help audit our particle systems.

In the next chapter, we will look at integrating Blueprints and Niagara Systems. We have already done this at a smaller scale with our **Presence Detector** particle effect in *Chapter 7*. Now, we will do a deep dive into it.

11

Controlling Niagara Particles Using Blueprints

In this chapter, we are going to use all the things we learned in the previous chapters. We will create a moderately complex Niagara particle system, and in that system, we will have some User Exposed parameters. We will be creating a Blueprint Actor that will drive these User Exposed parameters so that we do not have to directly interact with the Niagara System. Once you are familiar with this workflow, you will be able to create blueprint assets containing Niagara particle systems that can be tweaked without having to open the Niagara Editor.

The topics we'll cover in this chapter are as follows:

- Exploring the **User Parameters** module
- Calling Niagara User Exposed Settings from Blueprint Actors
- Tutorial – Modifying a Niagara System using Blueprint Actors

Technical requirements

Like with all the previous chapters, you will require Unreal Engine 5.1. Along with that, you will also need an image editor such as Affinity Photo or Photoshop.

You can download Affinity Photo from `https://affinity.serif.com/en-us/photo/`.

You can find the project we worked on in this book here:

`https://github.com/PacktPublishing/Build-Stunning-Real-time-VFX-with-Unreal-Engine-5`

Exploring the User Parameters module

We saw how to add the Niagara particle system to blueprints in *Chapter 4*. In this chapter, we will go a bit deeper and control Niagara parameters through blueprints. This allows us to create assets containing

Niagara particle systems that can be tweaked by users who are now familiar with Niagara. Through a blueprint, we can expose properties that we expect to be tweaked by level designers and other artists not directly working with Niagara. We are adding a layer of abstraction on top of the Niagara interface to make tweaking the particle system property values easier for other users.

The first step to do this is to create what we call User Parameters in the Niagara particle system, which will be exposed to the blueprints.

Let's understand this process by creating one such parameter.

To do that, first, create a new Niagara system. Use the **Fountain** template when asked to choose the template. The Fountain emitter has a **SpawnRate** parameter, which determines the number of particles emitted by the emitter. We will expose this parameter through a blueprint. We cannot do that directly, so we have to have a User parameter as an intermediary.

Figure 11.1: Create a new Niagara System using the Fountain template

In addition to having the emitter overview nodes, a Niagara System has an additional **System overview** node. This is a blue-colored node. The parameters that we create will be available in the **User Parameters** module in the **System overview** node. If you were to click on the **User Parameters** module now, you would see no parameters as we have yet to add them.

Figure 11.2: The User Parameters module in the System overview node

To add a User parameter, click on the + sign on the right of the **User Exposed** section in the **Parameters** panel. This will open a menu that will allow you to select the type of parameter you want to create.

We plan to expose **SpawnRate** of the Fountain emitter. We know that **SpawnRate** is a float value. Therefore, we will choose the float type for the User parameter to be created, which can be found under **Make New | Common**.

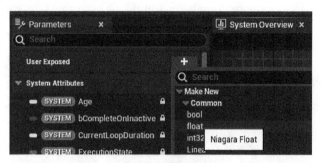

Figure 11.3: Creating a new float parameter under User Exposed

This will add the float parameter in the **Parameters** panel under the **User Exposed** section.

We know that it is a float by the color of the pill-shaped indicator next to it, which is green in color. The pill-shaped indicators are color coded depending on the parameter type. We are also informed that the parameter that we created is in the **User** namespace. Let us rename the parameter SpawnRate. We are naming it similar to the property that we intend to drive, which in this case is the **SpawnRate** property in the Fountain emitter. You can choose to name it differently.

Figure 11.4: Naming the new float parameter SpawnRate, under the USER namespace

Now, locate the **SpawnRate** property in the Fountain emitter. If you have modified the value (in this case, we modified it to 100.0), note it down for use later.

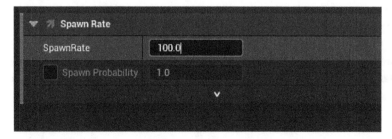

Figure 11.5: Check the SpawnRate value in the emitter SpawnRate module

Now, drag the User Exposed **User.SpawnRate** property from the **Parameters** panel into the **SpawnRate** property in the Fountain emitter properties, as shown in *Figure 11.6*. The property value box will be highlighted in blue dotted lines to indicate that the dragged property is compatible. Other properties that can accept a float value, such as **Spawn Probability**, will also be highlighted, but you can ignore them.

Figure 11.6: Drag the USER SpawnRate parameter into the SpawnRate property in the SpawnRate module

Release the **User.SpawnRate** property in the **SpawnRate** value box and you will now see a purple chain link indicator (*Figure 11.6*).

Figure 11.7: The purple chain link indicating that the SpawnRate property
is being driven by the USER SpawnRate parameter

This indicates that the **SpawnRate** property in the Fountain is now being driven by the **User.SpawnRate** property value.

We are now ready to create the blueprint that will let us modify the **User.SpawnRate** property outside of Niagara. As we will see in the upcoming tutorial, any property that we wish to modify outside Niagara has to be exposed similarly. Be aware that not all properties can be modified this way, but most of the common properties can. For more complex values and objects, it is recommended to use the **Niagara Data Interface**, which is an advanced functionality and has not been covered in this book.

Calling Niagara User Exposed settings from Blueprint Actors

Now that we have created a Niagara System, let us start with the blueprint. We will create the blueprint, connect it to the User Exposed parameters, and then expose those parameters as public variables for users to edit them:

1. Let us start off by creating an Actor Blueprint. Let us call it `BP_Fountain`.

Figure 11.8: Creating a new Blueprint Actor class and naming it BP_Fountain

2. We need to add a Niagara particle system component to this. You can choose to dynamically create one using blueprints or keep things simple and add it by clicking on the + **Add** button in the **Components** panel. Let us choose the simple way. Rename the component that we added `NiagaraFountain`.

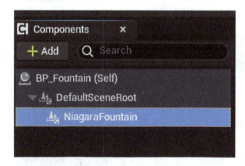

Figure 11.9: Add a Niagara particle system component and name it NiagaraFountain

3.   In the **Details** panel, select the **NewNiagaraSystem1** particle system we created in the previous section under the **Niagara System Asset** property. Our blueprint now has all the assets needed.

Figure 11.10: Selecting the NewNiagaraSystem1 particle system under the Niagara System Asset property

Let us now write the blueprint script required to connect the User Exposed parameters to public variables in the blueprint.

Since we expect users to tweak the Blueprint Actor properties, in **Editor** mode, our script will be written in the **Construction Script** tab. While you can write this code in the **Event Graph** connected to **Event Begin Play**, if you do so, it will not run while you are in **Editor** mode and will only run when you press **Play**.

We are aware that the **SpawnRate** and **User.SpawnRate** parameters are of the **float** type. So, the appropriate node that we can use to modify **User.SpawnRate** is the **Set Niagara Variable (Float)** node. You have different options depending on the type of parameters you are changing in the Niagara particle system, such as **Set Niagara Variable (Integer)** and **Set Niagara Variable (Linear Color)**. You can add the **Set Niagara Variable (Float)** node by dragging the **Niagara Fountain** component into the **Construction Script** editing area and then dragging a line from the **Niagara Fountain** component. This will pop up a context-sensitive menu. Search for Niagara Variable and one of the items that will show up is **Set Niagara Variable (Float)**. Select the **Set Niagara Variable (Float)** item, which will add it as a node to the graph.

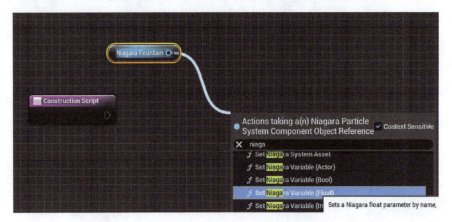

Figure 11.11: Adding the Set Niagara Variable (Float) node

Connect the execution line (the white line) of this node to the **Construction Script** node, as shown in *Figure 11.12*.

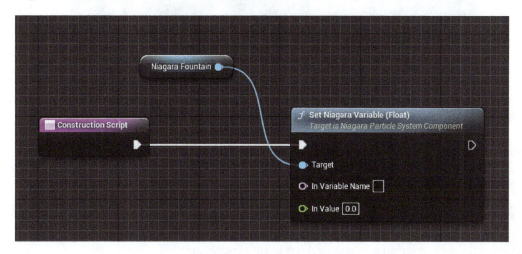

Figure 11.12: Connect the execution pins between the Construction
Script node and Set Niagara Variable (Float)

In the Niagara System, go to the **User Exposed** parameters area, right-click on the **User.SpawnRate** parameter, and choose **Copy Reference**. We will be pasting this reference in the **Set Niagara Variable (Float)** blueprint node we just added.

Figure 11.13: Right-click on USER SpawnRate and select Copy Reference from the pop-up menu

Back in the blueprint, paste the reference into the **In Variable Name** text box. The text **User.SpawnRate** will be pasted into the box. The blueprint node will now be setting the **User.SpawnRate** value using **In Value**. We need to make this **In Value** public so that users can access it directly from the **Details** panel in the main editor. **In Value** cannot directly be made public; we need to promote it to a variable first. To do this, let us promote **In Value** to a variable by right-clicking on it and choosing **Promote to Variable** from the pop-up menu.

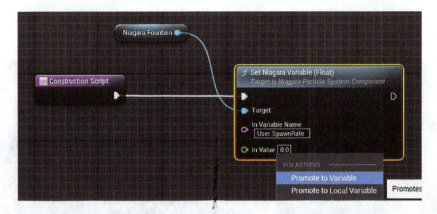

Figure 11.14: Pasting the User.Reference text into In Variable Name and promoting In Value to a variable

Let us rename the variable `ParticleSpawnRate`. We chose this name as it will be easier for someone editing the values in the blueprint to understand that it is a particle property. As with other variables, the name does not matter, so feel free to choose any name.

Figure 11.15: Name the promoted variable ParticleSpawnRate

Our Construction Graph should now look like *Figure 11.16*.

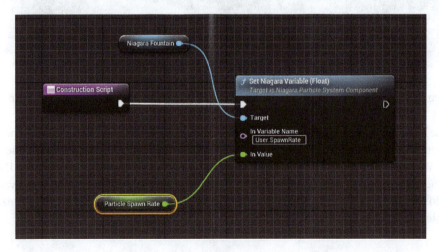

Figure 11.16: The finished Construction Script

Let us make the **ParticleSpawnRate** variable public by clicking on the closed-eye icon on the right (see *Figure 11.17*) where it is listed in the **VARIABLES** section of the **My Blueprint** panel. The eye icon will change to an open eye, indicating that the variable is now a public variable.

Figure 11.17: Make the ParticleSpawnRate float variable public by clicking
on the closed-eye icon to change it to an open-eye icon

Click on the **Compile** blueprint in the toolbar at the top.

Figure 11.18: Compile the blueprint script

For the **ParticleSpawnRate** variable, set the default value to any number. In this case, we are setting it to 90.

Figure 11.19: Set the default value of the public variable
ParticleSpawnRate to 90 or any number of your choice

Save and compile the blueprint again. Drag the blueprint into the level. In the **Details** panel of the blueprint object, you should be able to see the exposed **Particle Spawn Rate** property in the **Default** section.

Figure 11.20: The public variable ParticleSpawnRate seen in the Details
panel of the blueprint in the main editor window

The blueprint should look something like seen in *Figure 11.21* in your level. You will see the Fountain particle system active and emitting particles at a SpawnRate of 90.

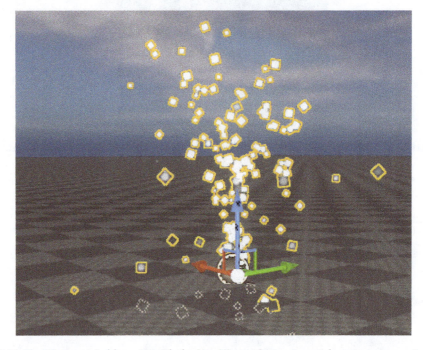

Figure 11.21: The BP_Fountain blueprint with the NewNiagaraSystem1 particle system as seen in the level

In the **Details** panel of the blueprint, change the exposed **Particle Spawn Rate** variable to a higher value, as shown in the following screenshot.

Figure 11.22: Tweak the ParticleSpawnRate value in the Details panel of the BP_Fountain Blueprint Actor

You will see the particle system in the blueprint react to this change, and the number of particles being emitted increases. The blueprint variable is now affecting the **SpawnRate** parameter in the Fountain emitter.

Figure 11.23: Our particle system reacting to the changes to the public variable ParticleSpawnRate

Now that we have understood how to control Niagara System parameters through blueprints, we will take you through a tutorial that uses this workflow to develop a particle system in which we have fire emitting from a logo. In the example we'll use in the tutorial, we will add a few properties that will let us control the particle system properties from a blueprint.

Tutorial – modifying a Niagara System using Blueprint Actors

In this section, we will set up a blueprint containing a Niagara particle system. The Blueprint will have a texture of a logo and we will have the Niagara particle system emit fire through the shape of the logo (see *Figure 11.24*).

Figure 11.24: The final effect we will achieve by the end of this tutorial

To achieve a result akin to *Figure 11.24*, with controls built into the asset, we need to change the logo, the density of the fire, and the color of the fire. We will need the help of a blueprint like the one shown in *Figure 11.25* to do this. This schematic shows the organizational structure of the Blueprint Actor containing the Niagara particle system and the names of the controls we will build into the blueprint.

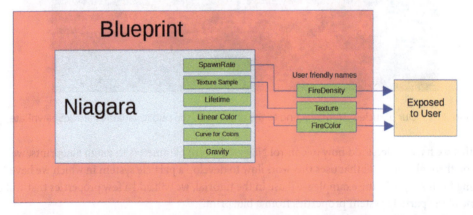

Figure 11.25: A schematic representation of the blueprint we will develop in this tutorial

We will expose parameters to change the amount of fire, the color of the fire, and the logo itself. Our Niagara system contained in the blueprint will react to the changes. Ensure that **Starter Content** is loaded for the project as we will be using some of the assets from it. If you do not have **Starter Content** loaded in your project, you can load it by following these steps:

1. Click on the + **Add** button at the top of the Content Browser.
2. Choose the **Add Feature or Content Pack…** option in the menu that pops up.

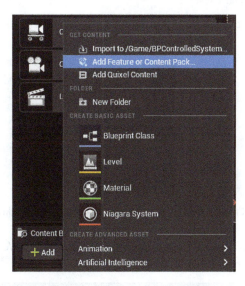

Figure 11.26: Add Starter Content by clicking on the + Add button
and selecting Add Feature or Content Pack…

3. A dialog box will pop up. Choose the **Content** tab at the top of the dialog box.
4. From the screen that appears, choose the **Starter Content** option, and then click on **Add to Project**. A new **Starter Content** folder will be created in your project's **Content** folder.

Figure 11.27: Select Add to Project on this screen to add Starter Content to your project

With **Starter Content** loaded in your project, you will have access to a bunch of assets that we will use in the Niagara System that we will create in this chapter.

Creating the Niagara System

Let's start by creating the Niagara System:

1. As usual, we will choose the **Fountain** template. Let us rename our Niagara System `NS_PacktFireLogo`.

Figure 11.28: Create a new particle system and name it NS_PacktFireLogo

2. Double-click on the Niagara System to open the Niagara editor. We will see some overview nodes (as seen in *Figure 11.29*).

Figure 11.29: The particle system nodes at the start as derived from the Fountain template

Now, let us make a few modifications to the emitter node to help us achieve the effect we want:

1. The first thing we will do is disable the **Add Velocity** module as we will not be needing it. Feel free to delete it if you want.

Figure 11.30: Disable the Add Velocity module

2. Since we want the flames to rise, we will be modifying the **Gravity** module values to make the particles move upward with acceleration.

   We will do this by changing the **Z** value of **Gravity** to 400. The **Gravity** module is essentially a force module, and it does not always need to point downward. We can use it as we like.

Figure 11.31: Modifying the Gravity Force properties to make the particles move upward like a flame

3. Next, we will go to the **Initialize Particle** section in the **Fountain** overview node and change the values of **Lifetime Min** to 0.2 and **Lifetime Max** to 0.5.

   We will be tweaking these values as we work our way through this exercise till we are happy with them.

Figure 11.32: Modifying the Lifetime properties as per our creative needs

Our Fountain particle system should now look something like *Figure 11.33*.

Figure 11.33: The Fountain emitter after the modifications

Our desired effect is to have a mesh with the logo texture applied and have the particles emit only from the white areas of the texture. To do this, we will have to make sure that the particles first emit from the whole mesh by sampling the mesh. We will then use the texture to mask the particles in a way that they appear to emit only from the white areas of the texture.

To use the texture as a mask, we will have to sample the texture. Sampling the texture using the CPU is a very expensive process, so we will use the GPU for the sampling.

Sampling the mesh

Let us start by sampling a static mesh:

1. We can do this by adding the **Sample Static Mesh** module to the **Particle Spawn** group.

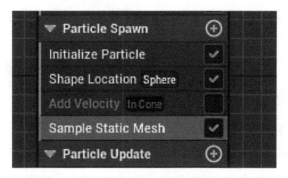

Figure 11.34: Add the Sample Static Mesh module

2. On adding this module, you will see a bunch of warnings pop up. These warnings appear because **Sample Static Mesh** is expecting a mesh to be assigned.

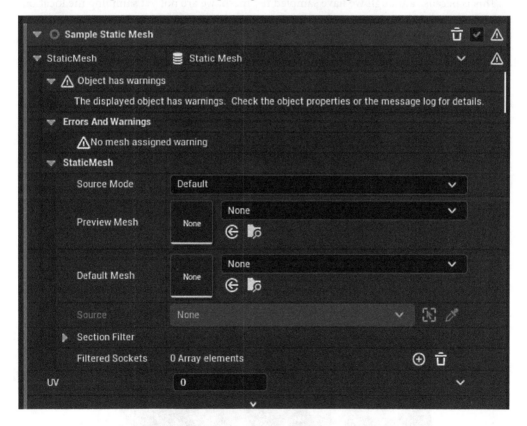

Figure 11.35: The Sample Static Mesh module warning us that no mesh has been assigned to sample from

3. Assign the **Shape_Plane** mesh, which is a part of **Starter Content**. There is no need to assign meshes to **Preview Mesh** as it is there as a fallback.

Figure 11.36: Assigning a Shape_Plane geometry to Default Mesh

If you follow the preceding steps, the warnings will disappear. However, you will see no difference in the particle system behavior in the preview window.

This is because although we have sampled the mesh, we are not yet sampling the location. Without sampling the location, the particle system does not have enough information to emit particles from the mesh.

4. To sample the location, add the **Static Mesh Location** module to the overview node under the **Particle Spawn** group.

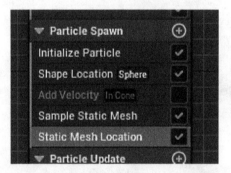

Figure 11.37: Add the Static Mesh Location module

5. To make sure we have a sufficient number of particles visible, set **SpawnRate** to 1000.0.

Figure 11.38: Set SpawnRate to 1000.0 for creative purposes

You should now see the particles emit from a square area corresponding to the area of the plane mesh that we are sampling.

Figure 11.39: Particles seen emitting from the whole surface of the mesh

In the next section, we want to set up the particle system so that it emits particles only from the white areas in the logo texture.

Sampling the texture

Let us prepare the logo first. You can create any black-and-white logo of your choice. We have created the Packt logo with a resolution of *1024 x 1024* and saved it as a .png file.

Figure 11.40 shows the logo texture we will use.

Figure 11.40: The Packt logo texture that we will use

1. Import this logo into Unreal as a texture by dragging the texture file from Windows Explorer into the Unreal Content Browser. We will name this logo `T_Packtlogo`.

Figure 11.41: The Packt logo texture imported into the Content Browser and named T_Packtlogo

2. We will now be sampling this texture to read the texture information into our particle system. We will be using this information to determine which area of the logo should be emitting particles. To sample the texture, add the **Sample Texture** module to our emitter node under the **Particle Spawn** group.

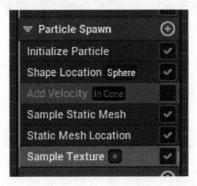

Figure 11.42: Adding the Sample Texture module in the Particle Spawn group

We will see a bunch of errors pop up again. These error messages inform us that the **Sample Texture** module does not work on CPU sims. Even if this were to work on a CPU, it would be very expensive, and texture sampling is best done on a GPU.

Figure 11.43: Sample texture module showing error messages
informing us that the module will not work on CPU sims

3. To convert our emitter to GPU, let us go to the **Emitter Properties** section and change **Sim Target** to **GPUComputeSim**. We may get a **Missing fixed bounds** message.

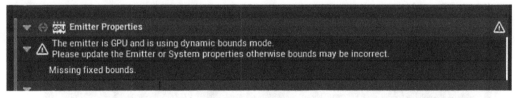

Figure 11.44: Emitter Properties module informing us that we are missing fixed bounds

4.   Select the **Fixed** option in **Calculate Bounds Mode** and this error message will disappear.

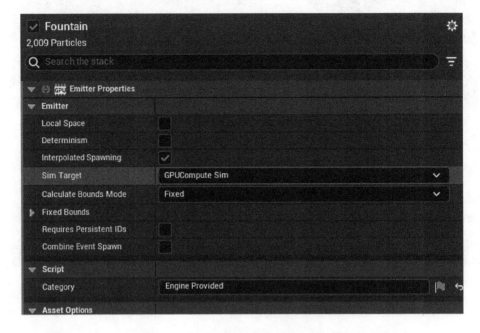

Figure 11.45: Set Calculate Bounds Mode to Fixed

In some cases, you might need to expand the **Fixed Bounds** property and adjust the dimensions of **Fixed Bounds**.

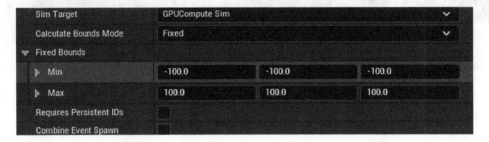

Figure 11.46: You may need to modify the Fixed Bounds Min and Max
values in some cases, but here, we are not changing them

We do not need to do this here. If you are wondering why we do not need to set the bounds in **CPUSim**, it is because when the emitter is a **CPUSim**, the bounds are created dynamically per frame.

The bounding box is used by the game engine to determine frustum culling. If the bounding box is not in the camera frustum, the engine does not render the simulation on screen.

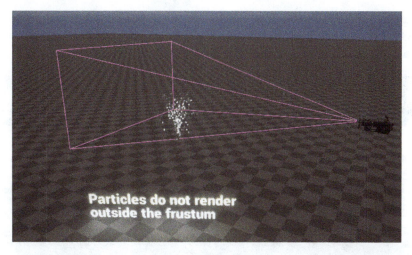

Figure 11.47: Unreal Engine culls objects outside of the camera frustum (shown in purple)

After setting **Sim Target** to **GPUCompute Sim**, the error messages for the **Sample Texture** module will vanish.

Now, it's time to select the Packt logo texture for the **Texture** property.

Under the **Texture** property, you will notice a **UV** property. This property reads the UV information to be applied to a texture for display. UV information helps the system understand how the texture is wrapped on the mesh. We need to read this UV from the mesh that we will be spawning the particles from.

Figure 11.48: Select T_Packtlogo for Texture

Now, it's time to select the Packt logo texture for the **Texture** property.

1. The **Sample Static Mesh** module writes out the UV information of the sampled static mesh
    to the **Particles.SampleStaticMesh.MeshUV** parameter. This parameter can be found in the
    Parameters panel. Drag this parameter to the **UV** property in **Sample Texture**.

Figure 11.49: The PARTICLES.SAMPLE STATIC MESH.MeshUV module found in the Parameters panel

As we have seen in earlier cases, the input boxes will be highlighted with a blue dotted line to
inform us that the dragged property is compatible.

Figure 11.50: Drag the PARTICLES.SAMPLE STATIC MESH.MeshUV module
into the UV property in the Sample Texture module

2. After dropping the parameter, the purple-colored chain icon will show up.

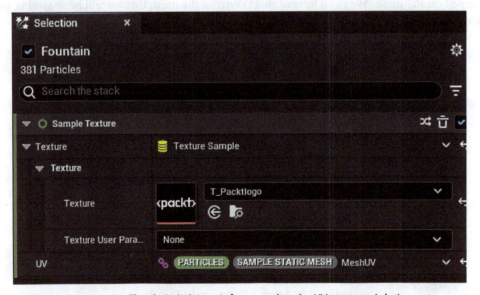

Figure 11.51: The chain link icon informs us that the UV property is being
driven by the PARTICLES.SAMPLE STATIC MESH.MeshUV parameter

The purple chain mentioned in *step 2* can be seen in *Figure 11.51*. This purple chain showing up means that the **UV** property under **Sample Texture** is now driven by the **Particles.SampleStaticMesh.MeshUV** parameter.

Emitting particles only from the white area of the logo

We now need to emit particles only from the white areas in the logo texture so that the particles appear to emit only from the logo text. To do that, we will, in fact, be killing all particles emitting from the areas that are not white. We do that by adding the **Kill Particles** module to the **Particle Spawn** section.

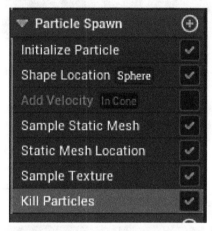

Figure 11.52: Add the Kill Particles module to the Particle Spawn
group in the NS_PacktFireLogo particle system

The **Kill Particles** module has a Boolean property, also called **Kill Particles**. If it is set to true, all emitted particles are killed. We will set up dynamic inputs to add conditions to the **Kill Particles** property. The conditions we set will enable us to have particles killed in only certain regions of the emitting mesh plane. This region will be defined by using a texture map. We will be comparing the value provided by the texture map with a threshold value. If the value supplied by the texture is greater than the threshold value, those areas of the texture will have the particles killed.

We already have the texture sampled as well as the mesh. We only need to provide a threshold value for the comparison. To do the comparison, we will add a dynamic input called **Set Bool by Float Comparison** to the **Kill Particles** Boolean.

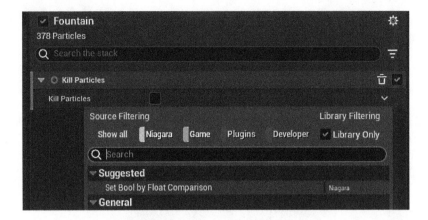

Figure 11.53: Add a Set Bool by Float Comparison dynamic parameter to the Kill Particles Boolean property

This will give us two float values, **A** and **B**, with a **Comparison Type** to compare the two floats. We will need to add another dynamic input to **A**.

This is because we will be doing a comparison on a float. We need to extract a float value from the texture. As an intermediate process, we will add a **Make Float from Linear Color** dynamic input to **A**. We will be extracting a single channel from this linear color value. In this case, let us choose the color red by selecting **R** from the **Channel** drop-down box under **A**.

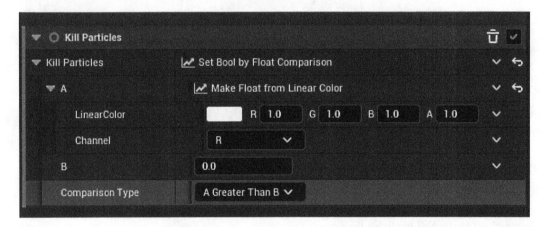

Figure 11.54: Add a Make Float from Linear Color dynamic input to A

Now, what we really want to compare is the logo texture in **A** and not the **LinearColor** property that **Make Float from Linear Color** provides against the threshold value in **B**. Because we added the **Sample Texture** module earlier, it will be writing out a parameter called **Particles.SampleTexture.**

SampledColor, which should be available in the **Parameters** panel. This **Particles.SampleTexture. SampledColor** contains information that the **Sample Texture** module has read from the logo texture.

Figure 11.55: The PARTICLES.SAMPLE TEXTURE.SampledColor parameter

Find and drag the **Particles.SampleTexture.SampledColor** parameter into the **LinearColor** slot.

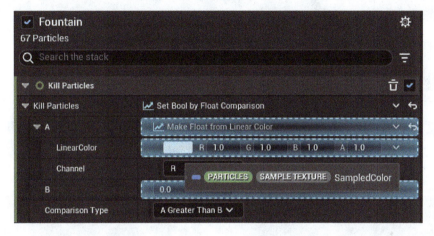

Figure 11.56: Drag the Particles.SampleTexture.SampledColor parameter into the LinearColor slot

The **LinearColor** slot will now have a purple-colored chain link icon (as seen in *Figure 11.57*) indicating that the **LinearColor** values are driven by the **Sampled Color** from the texture. Change **Comparison Type** to **A Equal To B** and set **B** to 0.0. Now, the Boolean value will be true when the value on the texture map is 0, that is, when the texture map is black.

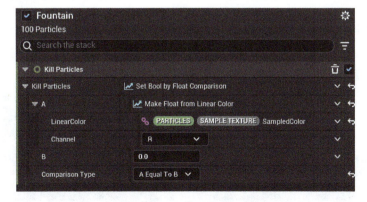

Figure 11.57: The chain link icon indicating that the LinearColor values
are driven by PARTICLES SAMPLE TEXTURE SampledColor

Your particle system should now be emitting particles from the white part of the texture. If the shape of the logo cannot be clearly seen, feel free to tweak the **Sprite Size** and **Lifetime** values of the particles. In our case, the values of 1.0 and 4.0 for **Uniform Sprite Size Min** and **Uniform Sprite Size Max** work.

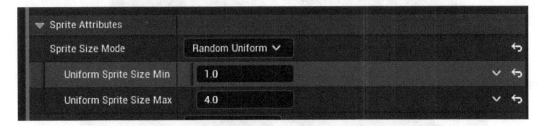

Figure 11.58: Setting the Uniform Sprite Size Min and Uniform Sprite Size Max values

For better readability of the logo, we have also tweaked **Lifetime Min** and **Lifetime Max** to 0.1 and 0.2, respectively.

Figure 11.59: Setting the Lifetime Min and Lifetime Max values for creative purposes

Our particle system should now look as in *Figure 11.60*.

Figure 11.60: The resultant particle behavior because of all the changes till now

While this particle system looks cool, we want the particles to look like fire. For this, we will need to do two things. First, we will need the particles to change color during the lifetime of the particle, and second, we will need an animated flame sprite.

To get the flame color to change during the lifetime of the particle, go to the **Scale Color** module, which should already exist in the **Fountain** template. If for some reason it is not there, add the **Scale Color** module to the **Particle Update** section.

In the **Scale Color** module, choose the **RGBA Linear Color Curve** option for **Scale Mode**.

This will change the properties shown for the **Scale Color** module to show a color gradient.

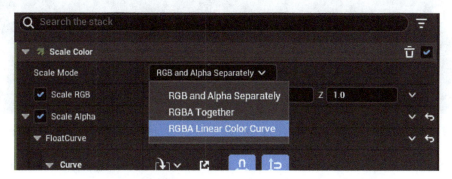

Figure 11.61: Choose the RGBA Linear Color Curve option for Scale Mode

This will add a color gradient to the properties. You can click on the top part of the gradient to add stops defining the color. At the bottom, you can define the alpha value (see *Figure 11.63* to see the gradient with the stops). The color gradient determines the color that the particle will have over its normalized age. A normalized age is when the age of a particle is denoted as a value between 0 and 1, with 0 being the age when the particle is spawned and 1 being the age of the particle when it dies, regardless of what the actual age of the particle may be in seconds.

At the beginning of the life of the particle, you can have the **V** value of the color be around `10.0`. This will make the particle glow.

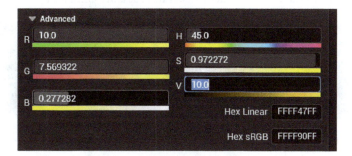

Figure 11.62: Setting the color on the first stop on the color gradient

Feel free to add color stops to **Linear Color Curve** as per your liking. **Linear Color Curve** should look something like *Figure 11.63*.

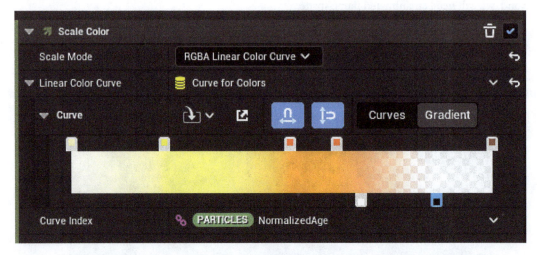

Figure 11.63: The color gradient with its final configuration of color (top) and alpha stops (bottom)

In the particle preview panel, you will see the particles changing their color throughout their life. We will now add an animated fire sprite to the particle sprite material.

Adding animated sprites using SubUV to the particles to have a flame animation

If you have **Starter Content** in your project, you should find the **M_Fire_SubUV** material by searching for it in the Content Browser. If **Starter Content** has not been added to your project, feel free to add it using the **Add** button at the top of the **Content Browser** panel.

In **Render** > **Sprite Renderer** > **Sprite Rendering**, select the **Material M_Fire_SubUV** material. This material has a sprite sheet with 36 images of a fire animation arranged in a *6 x 6* row-column structure. We will need to animate the SubUV to render these frames as a fire animation.

Figure 11.64: The M_Fire_SubUV material that will be used for the flame animation, by animating SubUV

Under **Sprite Renderer**, find the **Sub UV** section where we need to set **Sub Image Size** to 6 . 0 and 6 . 0, respectively, to correspond to the number of rows and columns in the sprite sheet.

Figure 11.65: As the sprite sheet is 6 x 6, set the Sub UV Sub Image Size values to 6.0 and 6.0

To trigger the animation of this sprite sheet, we need to add the **Sub UVAnimation** module to the **Particle Update** section.

Figure 11.66: Add the Sub UVAnimation module to the Particle Update section

In the **Sub UVAnimation** module, set the **End Frame** value to 35, which corresponds to the number of frames in the sprite sheet, which is 36 frames, since we start counting the frames from **Start Frame** 0. You may wonder whether this is redundant information as we already defined the number of rows and columns under **Sub UV**, and perhaps it may seem that the value of 35 should be evident to the emitter node. But the **End Frame** value is useful when you have only 34 frames, for example, in a 6 x 6 grid on the sprite sheet. It also helps us to pick only a part of the sprite sheet. For example, we may choose to use only frames 12 to 30 in our animation.

Figure 11.67: Setting the start and end frames for SubUV Animation

The particle system should now make it look like there is fire emitting from the logo. Under the **Sprite Attributes** section, we add 5 . 0 as the value of **Uniform Sprite Size Min** and 7 . 0 for **Uniform Sprite Size Max**.

Figure 11.68: Setting the Uniform Sprite Size Min and Uniform Sprite
Size Max values to adjust for the new flame sprites

Here, we add 0 . 2 as the **Lifetime Min** value and 0 . 5 as the **Lifetime Max** value. While the sprite size and the lifetime values shown in *Figure 11.68* and *Figure 11.69* are recommended, you can tweak them to your liking.

Figure 11.69: Setting the Lifetime Min and Lifetime Max values to adjust for the new flame sprites

After you have set the values of **Uniform Sprite Size Min** and **Uniform Sprite Size Max** to 5.0 and 7.0 and **Lifetime Min** and **Lifetime Max** to 0.2 and 0.5, respectively, set **SpawnRate** to 20000. The particle system in your preview window should look like *Figure 11.70*.

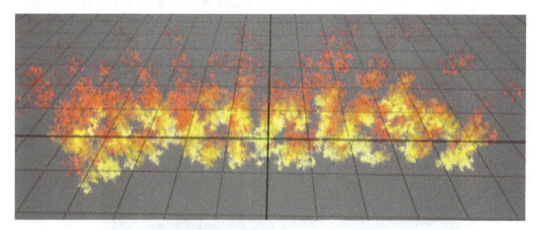

Figure 11.70: The particle system after the tweaks and the new animating fire sprites

Visually, our particle system is now ready.

Connecting User Exposed parameters to module parameters

We now need to add User Parameters to be able to drive them through blueprints.

We will be adding three User Parameters (as seen in *Figure 11.71*):

- **User.Texture** of type **Texture Sample** type to let us modify the logo texture from which the fire particles are emitted

- **User.FireDensity** of type **Float** to let us modify the number of fire particles

- **User.FireColor** of type **LinearColor** type to let us change the color of the fire particles

Figure 11.71: Adding the USER FireColor, USER FireDensity, and USER Texture User Parameters

This is how you can add the User Parameters:

1. We will be driving the appropriate properties in our emitter by these user created properties as we saw in the *Calling Niagara User Exposed Settings from Blueprint Actors* section at the beginning of this chapter. Drag the **User.Texture** property into the **Texture Sample** slot in the **Sample Texture** module.

Figure 11.72: Drag USER.Texture into the Texture Sample slot

This will replace the **Texture** slot with a link to **User.Texture** indicated with a purple chain link. Texture Sample will now be read from the **USER.Texture** parameter.

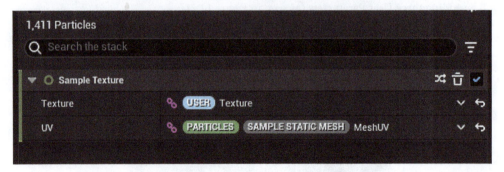

Figure 11.73: The familiar purple chain link indicating that the Texture
is being sampled from the USER.Texture parameter

2. Similarly, drag **User.FireDensity** and drop it into the **SpawnRate** slot, as shown in *Figure 11.74*.

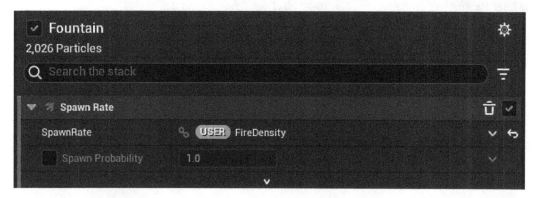

Figure 11.74: The purple chain link indicating that SpawnRate is driven by the USER.FireDensity parameter

3. Finally, drag **User.FireColor** and drop it into the **Color** slot in the **Initialize Particle** section.

Figure 11.75: The purple chain link indicating that the particle color
is being driven by the USER.FireColor parameter

All of our three User Parameters are now connected. Let us verify these connections by clicking on the **User Parameters** module in the **NS_PacktFireLogo System** module.

Figure 11.76: Click on the User Parameters module in the System node

In the **User Parameters** panel, we should see the three User Parameters that we created. We can set some default values here for the three parameters, as shown in *Figure 11.77*.

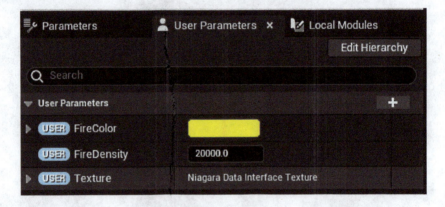

Figure 11.77: Setting the default values for User Parameters

We are now ready to move on to creating the blueprint.

Creating the Blueprint Actor

With our Niagara particle system done, let us start creating the Blueprint Actor class. We will name this Blueprint Actor BP_LogoFire.

Figure 11.78: Creating the Blueprint Actor class

Next, let's set **Niagara System Asset** to **NS_PacktFireLogo**:

1. Add a **NiagaraParticleSystem Component** to the actor and name it `NiagaraFire`. Of course, any name is fine.

Figure 11.79: Adding NiagaraParticleSystem Component

2. In the **Details** panel for the Niagara component, select the **NS_PacktFireLogo** option of the **Niagara System Asset**.

Figure 11.80: Set Niagara System Asset to NS_PacktFireLogo

And done!

Figure 11.81: Override Parameters in our Niagara Component Details panel

You will also see the User Parameters with their default values in the **Details** panel of the Niagara component. We can use this to modify the behavior of the Niagara component if needed, but we will be going a bit further and creating public variables in the blueprint to drive the User Parameters.

Creating the material for the plane geometry

We also need to make a material that we will apply to a plane shape. The plane shape will be a part of the Blueprint Actor and will be placed just under the particle system. This way, we can see the logo on the plane shape along with the particle system. This will enhance the readability of the logo:

1. Create a new material asset and name it M_packtlogo_Mat.

2. Let us add a **Texture Sample** node and a **Constant3Vector** node. Multiply these two nodes and connect them to the **Base Color** and **Emissive Color** pins of the result node.

    Since we plan to change the texture and the color properties in the material through blueprints, let us convert the **Texture Sample** and **Constant3Vector** color nodes in the material graph into parameters by right-clicking on the nodes and choosing **Convert to Parameter**. We will name them TextureUsed and LogoColor. The material node structure should look as in *Figure 11.82*. Set the values of the **LogoColor** parameter to **R** = 5.0, **G** = 2.5, **B** = 0, and **A** = 1.

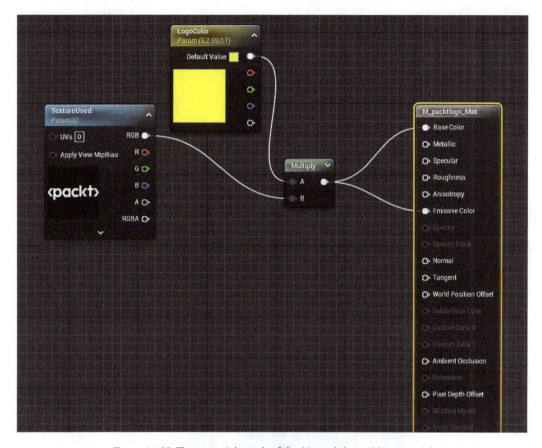

Figure 11.82: The material graph of the M_packtlogo_Mat material

The resultant material previewed on a box mesh should look as in *Figure 11.83*. Note that we have set the RGB color values to be greater than 1. This makes the logo glow.

Figure 11.83: The M_packtlogo_Mat material as seen in the preview window

3. Save the material, and this action should also update the material's thumbnail in the Content Browser.

Figure 11.84: The M_packtlogo_Mat thumbnail updated in the Content Browser

4. We will now add a **Plane** component to our blueprint by clicking on the **Add** button in the **Components** panel. Our **M_packtlogo_Mat** logo material will be added to this **Plane** component for better readability of the logo.

Figure 11.85: Add a Plane component to the Blueprint Actor

5. We will apply the **M_packtlogo_Mat** material that we just created to the plane. To do this, select the **Plane** component, and in the **Details** panel, choose the **M_packtlogo_mat** option for **Element 0**.

Figure 11.86: Apply the M_packtlogo_Mat material to the Plane component

6.   For aesthetic reasons, set the **Location** of the **Plane** component to 0.0, 0.0, -3.0. This way, the fire emitted by our Niagara system will be placed such that it will appear to be emitting from the Plane.

Figure 11.87: Setting the location of the Plane component just below the fire particle system

7.   If you haven't already dragged the blueprint into the viewport, do so now so that you can see it in action.

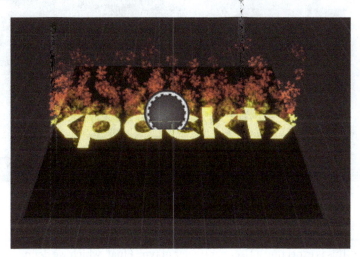

Figure 11.88: The blueprint as it appears in the level in the main editor viewport

We should be able to see the Niagara particle system and the textured plane together, as shown in *Figure 11.88*.

Connecting the blueprint public variables to the Niagara User-Exposed Parameters

We will now add public variables to the blueprint so that we can modify the Niagara particles' behavior directly from the **Details** panel of the main window.

Let us begin by exposing the **FireDensity** parameter.

Start by dragging the **Niagara Fire** component in the blueprint into the **Construction Script** graph. From the **Niagara Fire** node, drag a line and search for **Set Niagara Variable (Float)**. We will use **Set Niagara Variable (Float)** because the **FireDensity** parameter that we wish to connect this to is a float parameter.

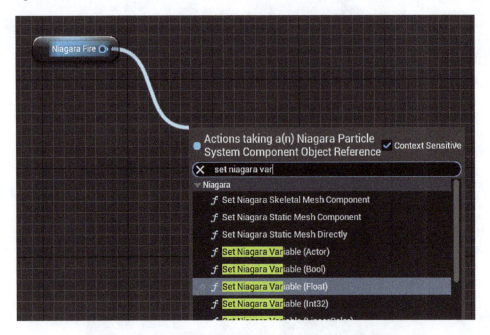

Figure 11.89: Adding the Set Niagara Variable (Float) node

Once the **Set Niagara Variable (Float)** node is added, there is an empty entry box in the node labeled **In Variable Name**. This is expecting a User parameter of type **Float**, which we defined in our Niagara system. In this case, we need to connect it to the **User.FireDensity** User parameter, which we have defined in the **NS_PacktFireLogo** particle system.

Go to your Niagara particle system, and in the **Niagara System Parameters** panel, right-click on the **User.FireDensity** parameter and choose the **Copy Reference** option.

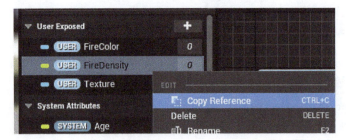

Figure 11.90: Right-click on USER FireDensity and select Copy Reference from the right-click menu

Paste the contents into the **In Variable Name** input box in **Set Niagara Variable (Float)**. It should paste the **User.FireDensity** text into the box.

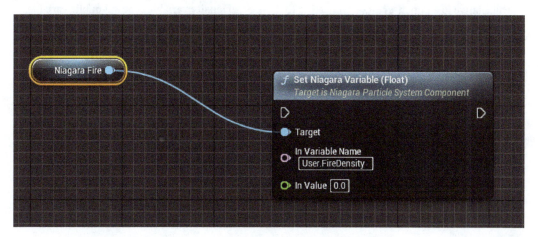

Figure 11.91: Paste the Copy Reference content into In Variable Name

Connect this node to the **Construction Script** node to see its effect in the viewport. You should now see the effects of the **Set Niagara Variable (Float)** node in your blueprint instance that you dragged into the level. Your fire particles should have disappeared. That is because **In Value** is set to **0.0**. You can play around with the value of **In Value** to see whether the blueprint is working.

We are going to expose this **In Value** as a public variable. To do that, right-click on **In Value** and choose **Promote to Variable** from the pop-up menu.

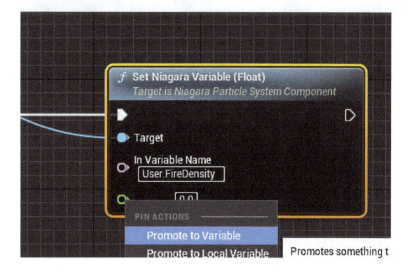

Figure 11.92: Promote the In Value pin to a variable. Call it FireDensity

This will create a new float variable. Rename the float variable `FireDensity`. Now, click on the closed-eye icon on the right of the **FireDensity** variable entry in the **VARIABLES** section. The closed-eye icon will change to an open-eye icon indicating that the **FireDensity** variable is now a public variable.

Public variables are shown in the **Details** panel when a Blueprint Actor is chosen in the level editor window.

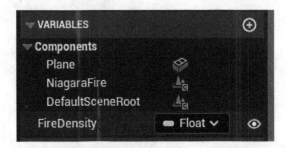

Figure 11.93: Make the FireDensity variable public by clicking on the eye icon next to it

We need to set the **Default Value** of the **Fire Density** variable. You can set the default value in the **Details** panel on the right side of the blueprint editor. Let us set the **Fire Density** value to 20000.0. This value is a creative choice, so feel free to tweak it as per your liking.

Figure 11.94: Set Fire Density to 20000.0

Let us now create similar public variables for the other User Parameters that we had defined in the Niagara system. The next parameter we will set is **User.Texture**. For that, drag a line from the **Niagara Fire** node and create a **Set Texture Object** node. As we did earlier, using **Copy Reference**, copy the reference from the **User Parameters User.Texture** into the **Set Texture Object Node's Override Name** pin. **Promote to variable** the **Texture** pin and create a **Texture** variable (see *Figure 11.95*). Then, make the **Texture** variable public.

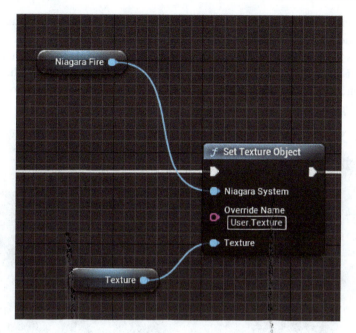

Figure 11.95: The Set Texture Object node in its final state

Set the default value of the **Texture** variable to the **T_Packtlogo** texture we imported earlier.

Figure 11.96: Set the default value of the Texture variable to T_Packtlogo

We will also be using this public variable Texture to set the **TextureUsed** parameter in **M_packtlogo_Mat**. This will make the texture on the Plane change in sync along with the texture supplied to the particle system.

To set up this functionality, drag the **Plane** component into the **Construction Script** window. From the **Plane** node, drag and create a new node called **Create Dynamic Material Instance**. The return value of this node will enable us to change the parameters in the material. In the **Create Dynamic Material Instance** node, select the **M_packtlogo_Mat** material for **Source Material**.

Drag a line from **Return Value** and create a node called **Set Texture Parameter Value**. Through this node, we will set the value of parameters we defined earlier in **M_packtlogo_Mat**. The parameter we will be setting is the **TextureUsed** parameter. Type the exact name of the parameter as it is in the **M_packtlogo_Mat** material nodes. To the **Value** pin of **Set Texture Parameter Value**, connect the same **Texture** public variable we created and connected to the **Set Texture Object** node. Your nodes should look as in *Figure 11.97*.

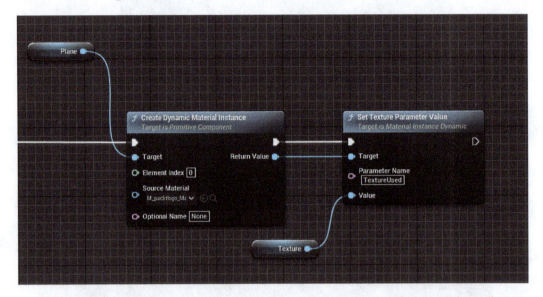

Figure 11.97: Add Create Dynamic Material Instance and the Set Texture Parameter Value node

Let us now connect the third User parameter we defined in the Niagara system, named **User.FireColor**, which will change the tint of color applied to the fire sprite. The **User.FireColor** variable is of type **LinearColor**. The process is the same as the one we used for the previous two parameters.

Drag the **Niagara Fire** component into the **Construction Script** graph. From the **Niagara Fire** node, drag a line and create the **Set Niagara Variable (Linear Color)** node. From the **User Parameters** panel in our Niagara System, use **Copy Reference** on **User. FireColor** and paste it into the **Set Niagara Variable (LinearColor)** node.

Then, right-click on the **In Value** pin on the **Set Niagara Variable (LinearColor)** node and Promote to Variable **FireColor**. Like our previous variables, make the **FireColor** variable public by clicking on the closed-eye icon next to it in the **Variables** section in the **MyBlueprints** panel.

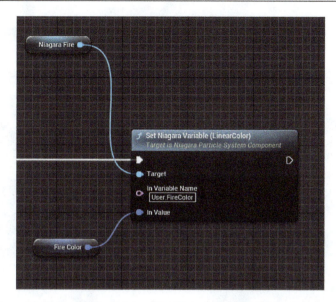

Figure 11.98: The finished state of Set Niagara Variable (LinearColor) node

We will be using this **FireColor** variable to also set the color of the logo. Obviously, we want the color of the logo and the fire to match. We therefore need to set the **LogoColor** parameter that we defined in our **M_packtlogo_Mat** material.

To do that, drag a line from **Return Value** on **Create Dynamic Material Instance** and create the **Set Vector Parameter Value** node. In the **Parameter Name** box of the **Set Vector Parameter Value** node, copy the parameter name exactly as it is in **M_packtlogo_Mat**. In this case, it is **LogoColor**. The node network should look as in *Figure 11.99*.

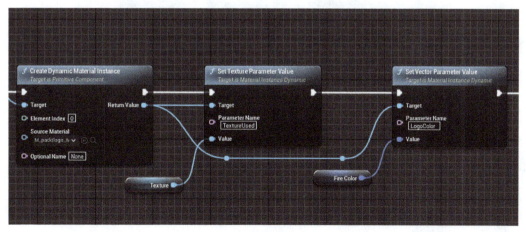

Figure 11.99: Adding the Set Vector Parameter Value node, setting Parameter Name to
LogoColor, and connecting the previously created FireColor variable to the Value pin

Let us set the default value of **Fire Color** to **R** = 5.0, **G** = 2.59, **B** = 0.0, and **A** = 1.0. This will give the logo and the fire a glowing yellow look.

Figure 11.100: Setting the color values for the Fire Color variable

In the **MyBlueprints** panel, the variables should look as shown in *Figure 11.101*.

Figure 11.101: The public variables in the BP_LogoFire Blueprint Actor

Double-check that they all have been made public and are of the proper variable type.

Organizing our public variables into a category

To make it evident to the user that these are custom properties and not part of the default properties shown in the **Details** panel, we will put them under a descriptive **Category** section name. We will call this category PacktAttributes.

Select each of the public variables in the blueprint, and in the details of that variable, type `PacktAttributes` under the **Category** entry. You will only need to type this out for the first time.

Figure 11.102: Setting the Category of our public variables to PacktAttributes, typing it out the first time

For the second variable, you should be able to select **PacktAttributes** from the drop-down menu for **Category**.

Figure 11.103: Setting the category of the remaining public variables by selecting
the previously created PacktAttributes category from the drop-down box

The variables will now be categorized under the **Packt Attributes** heading under the **Variables** section as well as in the **Details** section in the level editor.

Figure 11.104: The public variables now appearing under the Packt Attributes group

The node graph in the Construction Graph should look as shown in *Figure 11.105*.

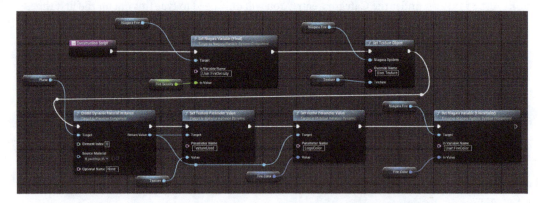

Figure 11.105: The full Construction Graph of the BP_LogoFire Blueprint Actor

The blueprint object in the level should look as shown in *Figure 11.106*.

Figure 11.106: The finished Blueprint Actor in the level editor viewport

Select the Blueprint Actor in the level, and in the **Details** panel, we will see the public variables under the **Packt Attributes** section.

Testing our Blueprint Actor

You can change the values in the **Packt Attributes** properties to modify the behavior of the particle system and also change the logo and its color.

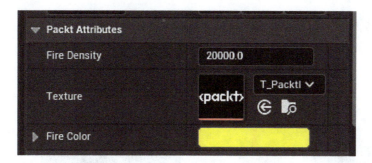

Figure 11.107: Changing the values of the public variable in the Blueprint Actor's Details panel

For example, changing **FireColor** to green will change the color of the logo and fire particles to green (as shown in *Figure 11.108*).

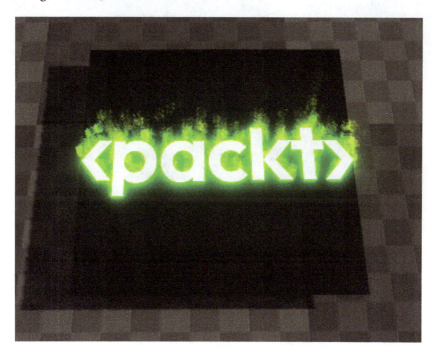

Figure 11.108: Changing the color of the fire to green by changing the FireColor public variable values

We should also be able to change the **Texture** property. Import any other logo into Unreal. The logo image should be black and white. In this case, I have imported the Unreal Engine logo and named it T_UE5Logo.

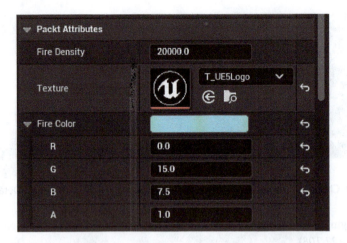

Figure 11.109: Changing the logo texture image

Now you should see the fire adjusting itself to emit only from the white areas of the Unreal logo.

Figure 11.110: The fire emitter particle's behavior changes according to the shape of the logo

Our tests of the Blueprint Actor are successful, and we are now able to modify the Fire Density, the Fire Color, and the Logo texture through public blueprints.

Summary

In this chapter, we learned how to create a fairly complex Niagara particle system and have it be a part of a Blueprint Actor. We then created blueprint scripts to enable us to change the properties of the Niagara actor through blueprints instead of directly editing the Niagara System properties. This way, we were able to expose only limited properties of the Niagara System to make it easy for users unfamiliar with Niagara to be able to work with our particle system.

We have come to the end of our journey of learning about Niagara particles. While we covered a lot of aspects of the Niagara particle system, there is a lot more to learn. Hopefully, we will meet in another book on Niagara to learn more.

Index

www.packtpub.com

Subscribe to our online digital library for full access to over 7,000 books and videos, as well as industry leading tools to help you plan your personal development and advance your career. For more information, please visit our website.

Why subscribe?

- Spend less time learning and more time coding with practical eBooks and Videos from over 4,000 industry professionals

- Improve your learning with Skill Plans built especially for you

- Get a free eBook or video every month

- Fully searchable for easy access to vital information

- Copy and paste, print, and bookmark content

Did you know that Packt offers eBook versions of every book published, with PDF and ePub files available? You can upgrade to the eBook version at packtpub.com and as a print book customer, you are entitled to a discount on the eBook copy. Get in touch with us at customercare@packtpub.com for more details.

At www.packtpub.com, you can also read a collection of free technical articles, sign up for a range of free newsletters, and receive exclusive discounts and offers on Packt books and eBooks.

Other Books You May Enjoy

If you enjoyed this book, you may be interested in these other books by Packt:

Blueprints Visual Scripting for Unreal Engine 5 - Third Edition

Marcos Romero, Brenden Sewell

ISBN: 978-1-80181-158-3

- Understand programming concepts in Blueprints
- Create prototypes and iterate new game mechanics rapidly
- Build user interface elements and interactive menus
- Use advanced Blueprint nodes to manage the complexity of a game
- Explore all the features of the Blueprint editor, such as the Components tab, Viewport, and Event Graph
- Get to grips with OOP concepts and explore the Gameplay Framework
- Work with virtual reality development in UE Blueprint
- Implement procedural generation and create a product configurator

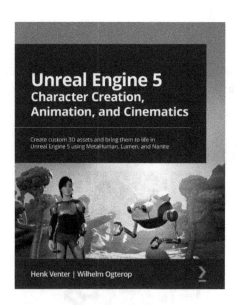

Unreal Engine 5 Character Creation, Animation, and Cinematics

Henk Venter, Wilhelm Ogterop

ISBN: 978-1-80181-244-3

- Create, customize, and use a MetaHuman in a cinematic scene in UE5
- Model and texture custom 3D assets for your movie using Blender and Quixel Mixer
- Use Nanite with Quixel Megascans assets to build 3D movie sets
- Rig and animate characters and 3D assets inside UE5 using Control Rig tools
- Combine your 3D assets in Sequencer, include the final effects, and render out a high-quality movie scene
- Light your 3D movie set using Lumen lighting in UE5

Packt is searching for authors like you

If you're interested in becoming an author for Packt, please visit `authors.packtpub.com` and apply today. We have worked with thousands of developers and tech professionals, just like you, to help them share their insight with the global tech community. You can make a general application, apply for a specific hot topic that we are recruiting an author for, or submit your own idea.

Share Your Thoughts

Hi!

I am Hrishikesh Andurlekar, author of *Build Stunning Real-time VFX with Unreal Engine 5*. I really hope you enjoyed reading this book and found it useful for increasing your productivity and efficiency.

It would really help me (and other potential readers!) if you could leave a review on Amazon sharing your thoughts on this book.

Go to the link below or scan the QR code to leave your review:

https://packt.link/r/1801072418

Your review will help us to understand what's worked well in this book, and what could be improved upon for future editions, so it really is appreciated.

Best wishes,

Hrishikesh Andurlekar

Download a free PDF copy of this book

Thanks for purchasing this book!

Do you like to read on the go but are unable to carry your print books everywhere?

Is your eBook purchase not compatible with the device of your choice?

Don't worry, now with every Packt book you get a DRM-free PDF version of that book at no cost.

Read anywhere, any place, on any device. Search, copy, and paste code from your favorite technical books directly into your application.

The perks don't stop there, you can get exclusive access to discounts, newsletters, and great free content in your inbox daily

Follow these simple steps to get the benefits:

1. Scan the QR code or visit the link below

https://packt.link/free-ebook/9781801072410

2. Submit your proof of purchase
3. That's it! We'll send your free PDF and other benefits to your email directly